NOR 4200
2nd edition

THE CHERNOBYL ACCIDENT AND ITS CONSEQUENCES

by

J H Gittus*

D Hicks***
P G Bonell**
P N Clough**
I H Dunbar**
M J Egan**
A N Hall**
M R Hayns***
W Nixon**
R S Bulloch****
D P Luckhurst****
A R Maccabee****
D J Edens*****

with a Foreword by the Lord Marshall of Goring, Kt, CBE, FRS
Chairman of the CEGB

* Corporate Headquarters, UKAEA
** Safety and Reliability Directorate, UKAEA
*** Harwell Laboratory, UKAEA
**** National Nuclear Corporation
***** Central Electricity Generating Board

United Kingdom Atomic Energy Authority
11 Charles II Street
LONDON SW1Y 4QP

April 1988

FOREWORD TO THE SECOND EDITION

By the Lord Marshall of Goring, Kt, CBE, FRS
Chairman of the CEGB

Since this report was first published in March 1987, more information and analysis has been released by the Russians through various international organisations. In addition, a number of reports from national and international bodies have been published providing a comprehensive coverage of international safety experts' views and details of the impact of the accident in other countries. The report has therefore been revised to reflect this new information and to provide a bibliography to the other major papers published elsewhere.

In terms of our understanding of the cause of the accident and of the physical and chemical processes that then occurred as the core overheated, there is little to add. All the international authorities agree on this. However, in addition to the design defects that were identified in our first report, it now seems likely that the magnitude of the accident was enhanced by a further phenomenon called 'positive scram'. This effect was first identified by a US Department of Energy team and has now been confirmed by the Russians. It is described in later sections of this report. It seems that in some unusual configurations the control rods of a Chernobyl type reactor do not always act to reduce and control the fission process. In these circumstances when they are first activated the initial effect is to <u>increase</u> the reactivity instead of decreasing it. Such a situation would have been unacceptable to the designers of a Western reactor but the Russian designers thought they could live with this strange phenomenon because of the low probability that this particular configuration of control rods would come about. But, as it happens, it did come about and it did come about in the Chernobyl incident and it now appears that this phenomenon enhanced the magnitude of the accident.

The Russians have also provided new information on radiation doses to people within the USSR. This has filled in certain gaps in the earlier information and has resolved a discrepancy in estimates of radiation dose that was highlighted in the first edition of this report. Extensive whole body monitoring of thousands of Russians has demonstrated that, as anticipated, the earlier estimates were too high. The latest Russian results compare favourably with our own earlier conclusions.

The other important development has been the proposal of a worldwide organisation of nuclear utilities to exchange information and experience on operational practices and to encourage emulation between them on good safety practice. I particularly welcome the commitment made by the Russians to this new organisation and their agreement that one of the four regional centres should be based in Moscow.

For the future, the Russians will order no more reactors of the RBMK design, but instead concentrate on their parallel PWR programme. They still plan a rapid expansion of their nuclear electricity generating capacity and intend to double capacity by 1990 compared with 1985 and increase capacity fivefold by the year 2000.

Marshall of Goring

Marshall of Goring

FOREWORD
By the Lord Marshall of Goring, Kt, CBE, FRS
Chairman of the CEGB

The Chernobyl accident was the most traumatic event in the entire history of civil nuclear power. For some months we could only speculate and guess about the causes of the accident but in August 1986, the IAEA organised an international conference in Vienna at which a group of Russian scientists and engineers made comprehensive presentations on what had happened. The presentation by the Russians was extremely frank and the nuclear industry throughout the world is now confident that it knows why the accident happened, how it happened and how it should have been prevented.

The Russian report is available to the general public, so is the report by the IAEA experts written immediately after the conference ended. But necessarily both those reports concentrate on what happened at Chernobyl. The purpose of this report is to do something different - namely to give an account of the Chernobyl incident from a UK point of view, comparing and contrasting safety, design and operational matters in Russia compared to UK practice.

This report was prepared by a team of people led by John Gittus of the UKAEA and has been reviewed for general accuracy by the UK nuclear industry as a whole. It represents the most authoritative and certainly the most complete response which the UK nuclear industry can give to the Chernobyl events at the present time. The subject is necessarily complex but we hope that the report will be widely read particularly by those with a technical interest in the accident and we would welcome comments or criticisms from other experts throughout the world.

Since this report was first published in March 1987, more information and analysis has been released by the Russians through various international organisations. In addition, a number of reports from national and international bodies have been published providing a comprehensive coverage of international safety experts' views and details of the impact of the accident in other countries. The report has therefore been revised to reflect this new information and to provide a bibliography to the other major papers published elsewhere.

In terms of our understanding of the cause of the accident and of the physical and chemical processes that then occurred as the core overheated, there is little to add. All the international authorities agree on this. It appears, however, that in addition to the shutdown system being too slow in its operation, it also had a serious design flaw that in the event could have significantly increased the magnitude of the power surge - the so-called "positive scram" effect. This effect, first identified by a US Department of Energy team, has now been confirmed by the Russians and is described in later sections of this report. To allow the potential for a positive scram effect would have been unthinkable for the designers of any UK reactor.

The Russians have also provided new information on radiation doses to people within the USSR. This has filled in certain gaps in the earlier information and has resolved a discrepancy in estimates of radiation dose that was highlighted in the first edition of this report. Extensive whole body monitoring of thousands of Russians has demonstrated that, as anticipated, the earlier estimates were too high. The latest Russian results compare favourably with our own earlier conclusions.

The Chernobyl events were so dramatic and involved such complexities that it is likely that different readers will draw different lessons from this report. I think it is worthwhile anticipating those differences in this foreword.

The reactor designer is likely to be both fascinated and distressed by the assessment of the Chernobyl reactors against British safety principles. The designer will draw reassurance from the fact, amply demonstrated in this report, that the Chernobyl design could not have been licensed or even considered for construction in the United Kingdom. The designer will conclude that there is nothing about reactor design that we can explicitly learn from the Chernobyl events because all the design questions which arise would have been anticipated by our own procedures. Indeed many of the design features were identified as adverse by a group of British Engineers who in 1976 studied the basic RBMK design of which Chernobyl is a development. But the designer's overwhelming reaction will be astonishment that the Russian designers disobeyed the fundamental first rule of reactor design which is to make sure that the reactor has a negative fast acting power coefficient.

The Russians neglected to ensure this for the Chernobyl reactor because at all powers below 20%, the fast acting power coefficient is positive. This is the factor that makes the design of Chernobyl unique. There is no other commercial nuclear reactor in the entire world that has a fast acting positive power coefficient. Fortunately the Russians have acknowledged and accepted this design fault and are retrofitting all their pressure tube reactors to eliminate this outstanding design "shortcoming" - to use the word the Russians have chosen themselves.

Although the Russians knew of this serious design "shortcoming", they did not take engineering steps to avoid it. They simply instructed their operators to avoid the dangerous regime. The instructions to the operators were very, very strict indeed. Legasov stressed that not even Mr Gorbachov had the power to overrule those instructions but the operators ignored them nevertheless. As Legasov said "Our designers made a colossal psychological mistake" in putting so much responsibility on the operators and relying upon the operators to obey those instructions literally and absolutely throughout the lifetime of the reactor. The operators did not obey their clear instructions. If they had obeyed the instructions, the accident would not have happened and at one level of understanding it is therefore possible to say that the Chernobyl accident was a consequence of operator error. But safety philosophy in the West is that operators should not be given the opportunity to make such devastating errors of judgement. Certainly reactors here in the UK have to be tolerant of operator error. For these reasons the main reaction of reactor operators in the UK will be reassurance that we do not depend solely upon them to guarantee safety.

However, that is only the first basic reaction of our British reactor operator. We have already anticipated operator errors and have designed UK reactors to be tolerant to those errors. But we have always assumed that operator errors in the UK would be sins of omission or commission or misunderstandings or errors of communication. The operator errors at

Chernobyl were not of that kind at all. They involved systematic, persistent and conscious violations of clearly stated safety rules. To us in the West, the sequence of reactor operator errors is incomprehensible. It was incomprehensible to Legasov also. Perhaps it came from supreme arrogance (we operators know better than the designers) or complete ignorance (Legasov says the operators had lost all sense of danger). More plausibly we can speculate that the operators as a matter of habit had broken the rules many, many times and got away with it so the safety rules no longer seemed relevant. Here in the UK, we must in all conscience address the question, "Could any of our operating teams get into the same state of mind?" At sometime in the future when nuclear power is an accepted and routine method of electricity generation could our operators become complacent or arrogant and deliberately and systematically flout operating rules? Could our system of safety reviews and independent assessment and inspection become so lax or so routine that it failed to identify and correct such a trend?

My own judgement is that the overriding importance of ensuring safety is so deeply engrained in the culture of the nuclear industry that this will not happen in the UK. But herein lies the challenge to our professional management: if we are so confident, will we become complacent? On the question of avoiding systematic and sustained operator error, therefore, we need to be confident that the situation is safe and yet remain constantly vigilant at the same time.

What lessons will our Government draw from this report? I think they are likely to note that there was no effective institutional restraint on the Russian system as a whole. All 15 pressure tube reactors of the Chernobyl design were potentially dangerous, all were operating and the Russian safety system gave no institutional alarm signals at all. Therefore, whatever the situation appears to be on paper, it seems clear that, in practice, the Russians did not have an effective independent safety inspectorate. I think a Government of the UK is likely to take reassurance from the fact that we do have an independent nuclear "watchdog" in the UK, namely the NII.

What will be the reaction of the British public? This of course is the most important reaction of all because without public acceptance we simply cannot have nuclear power. I think the public are likely to accept that the British reactors are designed better and operated better than Russian reactors. I think they will accept that an effective and independent nuclear inspectorate is an important additional safeguard. But it is inevitable that they will be left with an uncomfortable feeling that their safety is in the hands of other people and in a very broad sense the public will be correct to say that their safety does depend on the integrity and conscientiousness of the electricity utility operating the reactor; in this country primarily the CEGB. That is why we in the CEGB, with the close assistance of our colleagues in the rest of the nuclear industry, must strive to win the confidence and trust of the British people. There is only one way to do that: by continuing to manage our nuclear activities in an efficient, professional and above all safe manner and by giving as much information as possible to the British public in a language which is both correct and understandable. I think this report, though necessarily long and complex, is a good contribution to this latter aim.

For the future, the Russians will order no more reactors of the RBMK design, but instead concentrate on their parallel PWR programme. They still plan a rapid expansion of their nuclear electricity generating capacity and intend to double capacity by 1990 compared with 1985 and increase capacity fivefold by the year 2000.

Marshall of Goring

Marshall of Goring

SUMMARY

The accident to a Russian RBMK nuclear reactor which occurred on 26 April 1986, was due to three main design drawbacks:

1 the reactor had a positive void coefficient and, below 20% power, a positive power coefficient, which made it intrinsically unstable at low power;

2 the shutdown system was in the event inadequate and might in fact have exacerbated the accident rather than terminated it;

3 there were no physical controls to prevent the staff from operating the reactor in its unstable regime or with safeguard systems seriously disabled or degraded.

In their endeavours to perform an experiment, the operators violated their operating regulations and allowed the reactor to enter the unstable regime. In the UK HM Nuclear Installations Inspectorate's safety principles would not permit licensing of such an inherently unsafe design concept and would require automatic devices to prevent operation of any reactor in circumstances where its safety might be seriously compromised.

The accident was triggered by a turbogenerator experiment, when the reactor core contained water at just below the boiling point, but little steam. When the experiment began, half of the main coolant pumps slowed down and the flow reduction caused the water in the core to start boiling vigorously. The bubbles of steam that formed absorbed neutrons much less strongly than the water they displaced and the number of neutrons in the core started rising. This increased the power of the reactor, more steam was produced and even fewer neutrons were absorbed, a phenomenon known as "positive feedback". The reduction of neutron absorption caused the "excess reactivity" of the core to rise to the level where the chain reaction could be sustained by "prompt neutrons" alone, and the reactor became "prompt critical". The power surge caused the fuel to heat-up, melt and disintegrate. Fragments of fuel were ejected into the surrounding water, causing steam explosions that ruptured fuel channels and led to the pile cap being blown off. Radionuclides escaped into the atmosphere where the wind carried them to distant countries.

In the last few seconds of the incident, the operators realised that the power was rising and initiated the manual trip, but the shutdown capability had been degraded by maloperation of the control rods and the reactor did not shut down immediately. Moreover, the particular details of the system meant that there was a further power surge caused by its operation and, a few seconds after the trip button had been pressed, the unstable reactor went prompt critical.

The Russians are making a number of design changes to the RBMK reactors that appear to be sufficient to prevent such an accident happening again. These changes will address the design drawbacks so that:

i the reactor will be made more stable;

ii the shutdown system will be modified and made more rapid;

iii there will be additional shutdown signals to trip the reactor.

In addition, they wish to improve their understanding of the man-machine interface and will give additional training to the operators.

Thirty-one people, primarily those engaged in fighting fires caused by the accident, have died, mostly from the effects of very high levels of radiation exposure. Some two hundred others were diagnosed as suffering from acute radiation effects. 135,000 people were evacuated and an area out to about 30 kilometers is being decontaminated. Over the coming years, it is expected that regions of the contaminated land will be reclaimed for partial economic use.

Many millions of people in the USSR and Europe have received doses, for the most part small, of radiation as a result of the accident. In the vast majority of cases, however, the dose is no different to or smaller than the variation of the natural background radiation between different parts of the same country.

CONTENTS

EXECUTIVE SUMMARY

Following the accident at Chernobyl, there was naturally concern about its impact upon the status of nuclear power worldwide. In the early days, it was not easy for Western experts to respond to those concerns because very little information about the accident was available. However, the IAEA organised an international conference in Vienna in August, 1986, at which Russian scientists and engineers made comprehensive presentations on the accident. This Report draws extensively on that information.

Since this Report was first published in March 1987, more information and analysis has been released by the Russians, particularly in papers presented to the International Conference on Nuclear Power Performance and Safety held in Vienna, 28 September to 2 October 1987. In addition, a number of reports from National and International bodies have been published providing a comprehensive coverage of international nuclear safety experts' views, and details of the impact in other countries. This revision of the Report has been undertaken to reflect this new information, and to provide a bibliography to the other major papers which have been published elsewhere.

In terms of our physical understanding of the cause of the accident, and of the physical and chemical processes which then occurred as the core overheated, there is little to add. There is agreement between all the international authorities on this. One further mechanism which could have contributed to the positive power feedback has been identified which we did not cover. This is the effect known as 'positive scram'. A new section in Chapter A2.10 explains what this means and how it could have influenced the very early stage of the power excursion. This mechanism was first identified by a US Department of Energy team, but has now been confirmed by the Russians themselves in one of their recent papers [ref No 32 in the bibliography].

Much more information is now available on the radiological consequences both in the Soviet Union and other countries. It is therefore appropriate to update the relevant information in this Report. The new data has filled in certain gaps in the earlier information and has resolved a discrepancy in estimates of radiation doses in the USSR that was highlighted in the first edition of this Report. Extensive whole body monitoring of thousands of Russians has demonstrated that, as anticipated, the earlier estimates were too high. The latest Russian results compare favourably with our own earlier conclusions.

The Chernobyl accident was an event with implications for both the Russians and the rest of the world. Although the responsibility for the decisions and events that led up to the accident rests with the Russians, they command respect for the strenuous efforts they made to control the accident once it had occurred. All who are professionals in the nuclear business have been impressed with the bravery and the dedication of the operating engineers, firemen and men on the spot once the accident had occurred. The Russian authorities also moved with remarkable speed and efficiency to control the accident, mitigate its effects and direct the recovery process.

In addition, the international community must be grateful to the Russian authorities for the frankness with which they have described this event. All who went to the conference in Vienna came away with the strong conviction that the Russian engineers and technical experts had told everything that they knew themselves about the accident. There are still some points of detail to be resolved but there is general satisfaction that it is now known why the accident happened, how it happened and how it should have been prevented; this feeling has been strengthened by the further information now available.

As a result of this, the implications of the Chernobyl accident for the UK can now be addressed. It seems certain that a Chernobyl-type accident could not happen in the UK. UK safety rules first and foremost aim at the building of reactors that have intrinsic characteristics that provide inherent protection. Secondly, these natural defences are supplemented by engineered features to prevent, limit, terminate and mitigate any faults. Thirdly, the systems design must be tolerant to operator action - if the operator makes a mistake, the reactor shuts down. Fourthly, UK operators are highly educated and well trained, not just for routine operations but for unusual situations and accident situations and, fifth, the entire system is overseen by an independent Nuclear Inspectorate that can at any time, without hindrance or challenge, close down any licensed reactor.

The accident to an RBMK reactor at Chernobyl occurred because of design shortcomings, acknowledged by the Russians and leading to an instability so severe that it could not be corrected by inserting the automatic control rods. There were three main design shortcomings:

1 the reactor had a positive void coefficient and, below 20% power, a positive power coefficient, which made it intrinsically unstable at low power;

2 the shutdown system was in the event inadequate and might in fact have exacerbated the accident rather than terminated it;

3 there were no physical controls to prevent the staff from operating the reactor in its unstable regime or with safeguard systems seriously disabled or degraded.

Malpractice by the operators was a contributory cause and indeed they broke so many rules that one cannot help thinking that this was their regular habit. It is not credible that this was the one and only occasion on which they behaved in this manner. The Russians now acknowledge that the reactor designers should have envisaged such possibilities and guarded against them. They propose to make substantial changes to the design of RBMK reactors and will also give additional attention to the training of operators.

This Report on the Chernobyl accident and its consequences has three main objectives: to describe the plant and circumstances leading to the accident; to understand the development of the accident and its consequences; and to highlight the significant differences between Russian and UK nuclear safety standards revealed by analysis of the reactor design. To achieve these objectives, the Report is divided into Sections describing and discussing the Russian power reactor programme, the design of the Chernobyl reactor, its design defects, the proposed improvements to its safety,the development of the accident, the radiological source term to the environment and the environmental consequences. Comments on the lack of conformity of the Chernobyl reactor design with established UK safety principles are made at appropriate points in the text. The contents of the Sections are summarised below.

0.1 USSR Power Reactor Programme

The Russians have a large and expanding nuclear power programme. It is regarded as essential to their economy and currently provides about 15% of their power production. It is based upon two main types of reactor, the (RBMK) pressure tube reactor and (VVER) pressurized water reactors. The RBMK reactor is a direct cycle, boiling water, pressure tube, graphite-moderated reactor developed from the USSR's first nuclear power plant commissioned in 1954 at Obninsk. The concept is unique to the Soviet Union. Pressure tube reactors of the RBMK type have been operating in the USSR for over thirty years. At the end of 1985, fourteen large RBMK units were already in service, one was about to enter service and seven more were under construction. These range in output from 950 to 1450 MW(e) and are located on six sites. It is anticipated that the power generation at nuclear plants in the USSR will increase by a factor of 5 - 7 times over the 1985 value of 170TWh by the year 2000.

0.2 Description of the RBMK Reactors

The Chernobyl nuclear power station is located about 60 miles north of Kiev in the Ukraine on the Pripyat River not far from the town of Pripyat (pop 49,000). The site had four 1000MW(e) RBMK reactors operational and two more were under construction 1.5km away. The four reactors at Chernobyl were built in pairs, sharing common buildings and services. Construction of Units 3 and 4 started in 1975/76; Unit No 4 became operational during 1984.

The reactor core, 11.8m in diameter and 7m high excluding the reflectors, is built up from graphite blocks perforated by vertical channels each containing a zirconium alloy (Zr + 2.5% Nb) pressure tube 88mm external diameter and 4mm thick.

Each of the 1661 channels contains two fuel assemblies each 3640mm long held together by a central tie rod, suspended from a plug at the top of the channel.

The fuel assemblies consist of 18 pin clusters, each pin in the form of enriched (2%) uranium dioxide pellets encased in zirconium alloy tubing (13.6mm od x 0.825mm thick). The maximum power from any channel is 3.25MW.

The fuel is cooled by boiling light water at 70 bar pressure. The water enters the channel at 270°C and the steam quality of the exiting steam/water mixture is 14 weight per cent average (20% max).

Two separate similar coolant loops are provided. Each loop consists of two steam drums (to which the riser pipes from the fuel channels carry the steam/water mixture) and four primary circulating pumps (three are normally operational and one standby).

The dry steam from the steam drums passes to one of two 3000 rpm 500MW(e) turbine generators. The steam is condensed in surface condensers and the condensate is returned to the steam drums via a conventional feed train with electrically driven feed pumps.

About 5% of the energy of fission is dissipated in the graphite structure as a result of slowing down of neutrons and of gamma heating. This heat is transferred to the fuel channels by conduction and radiation via a series of 'piston-ring' type graphite rings which permit good thermal contact between the pressure tube and the graphite blocks whilst permitting some small dimensional changes. The maximum temperature of the graphite is about 700°C.

To improve the thermal contact and to prevent graphite oxidation, the graphite structure is enclosed in a thin walled steel jacket which contains a slowly circulating helium/nitrogen inert gas blanket.

Perhaps the most important characteristic of the RBMK reactor is that it has a "positive void coefficient". This can be explained in simple terms by recognising that if the power from the fuel increases or the flow of water decreases (or both) the amount of steam in the fuel channel increases and neutrons which would have been absorbed by the water pass more easily through the steam, causing additional fissions in the uranium fuel.

The term "positive void coefficient" then is the reactor physicists' phrase which expresses the fact that reducing coolant density results in an increase in neutron population and hence an increase of reactor power. However, as the power increases so too does the fuel temperature and this has the effect of reducing the neutron population ("negative fuel (Doppler) coefficient"). The net effect of a positive void coefficient and the negative fuel coefficient clearly depends on the power level.

In the RBMK reactor, at normal high power operating conditions the fuel temperature effect dominates and the net effect, termed the "power coefficient", is negative. However, below about 20% of full power, the power coefficient can become positive and the reactor becomes unstable. For this reason, operating below 20% power was restricted. This fundamental design shortcoming was the critical factor of the accident at Chernobyl. Another factor was the poor design and slow speed of response of the reactor control and shutdown system. The design of this system was seriously flawed as the insertion of graphite rods attached to the lower ends of the absorber rods could displace water and in the event caused an initial increase, rather than decrease in power. This is known as the 'positive scram' effect.

Channels for the control and shutdown rods and for the in-core flux instrumentation passed through vertical holes in the graphite blocks. Radial flux monitors were provided in over 100 channels and axial flux profiles were monitored in 12 channels.

The system for reactor control and protection used 211 solid absorber rods. It is now being modified in the light of the Chernobyl accident. The rods were divided functionally as follows:

a 163 manually operated rods of which 139 were control rods (RR) for radial power shaping and 24 were dedicated to emergency protection;

b 12 average power automatic regulation rods (AR1, AR2 and AR3);

c 12 local automatic regulation rods (LAR);

d 24 shortened absorber rods (USP) for axial flux profiling.

The manual control rods (RR), the automatically operated rods (AR) and the emergency shutdown rods (AZ) were distributed uniformly throughout the core in six groups of 30-36 rods. The control system included sub-systems for local automatic control (LAR) and local emergency protection (LAZ). All rods except the shortened absorber rods were withdrawn and inserted from above.

On emergency shutdown all the control rods except the shortened rods were motor driven at a speed of insertion of 0.4m/s. Full insertion took 15-20 seconds. The shortened absorber rods were introduced from below the core. The control rod channels were the same diameter as the fuel channels (88mm) and were cooled by a separate water circuit.

Reactor control is provided by several systems working together, which actuate different groups of control rods. These systems operate in different, but overlapping, power ranges to provide reactor control from subcritical to full power conditions. The power ranges and systems are:

1 from sub-critical to 0.5% power - manual;

2 from 0.5% to 10% power - low overall automatic regulation (AR3);

3 from 5% to 100% working range - overall automatic regulation (AR1 and AR2);

4 from 10% to 100% power - local automatic regulation (LAR).

The reactor is tripped only for a specific number of faults (eg loss of off-site (station) power, both turbines tripped, loss of three main circulating pumps, 50% loss of feedwater, low steam drum water level and high neutron flux).

For all other faults the reactor power is set back to some lower level consistent with the consequence of the fault on the reactor (eg loss of one circulating pump to 80% full power, trip of single turbine to 50% full power).

The RBMK reactors are designed to be refuelled at full load. The refuelling machine operates from a gantry running the length of the charge hall.

The high-pressure coolant system is housed in a series of compartments, some of which can act as the containment in the event of an accident. Separate compartments house the main coolant pumps, the coolant inlet headers and distribution pipework and the reactor vault.

The main coolant pump compartments are designed to withstand a pressure of 4.5 bar and are equipped with sealed electrical and mechanical penetrations and isolation valves. The compartments are connected via a surface condenser tunnel to two pressure suppression pools to condense the escaping steam and reduce the pressure. Excess pressure above the suppression pools can be relieved by venting into unaffected main coolant pump compartments.

The steam drums are housed in separate compartments on either side of the charge hall but these are not pressure tight compartments because of the large number of joints in the charge hall floor needed for refuelling which provide a leak path between the steam drum compartments and the charge hall.

The RBMK reactor is equipped with an emergency core cooling system that can feed either coolant loop and consists of:

1 a fast acting flooding system which automatically injects cold water to the damaged part of the reactor from two sets of gas-pressurized tanks holding enough water to cool the core for the first 100 seconds of a major loss of coolant accident. This system is supported with flow from the main feed pumps;

2 an active system of three pumps taking water from the condensate system. These pumps can be driven by three standby diesel generators which can be started within 15 seconds;

3 an active recirculating cooling system which consists of six pumps drawing water from the upper suppression pool through heat exchangers. This can also be driven by the diesels.

The emergency core cooling system is triggered by the coincidence of a high pressure signal from compartments containing the main coolant pipework and either a low water level signal from the steam drums or a signal indicating a decrease in circuit pressure drop. The latter identify the damaged half of the reactor. Subsystem 2 provides long term cooling for the undamaged loop and subsystem 3 provides long term cooling for the damaged loop.

0.3 NNC Safety Reservations

In 1975 active measures were in hand to promote relations between the USSR and the UK. Nuclear power was judged to be a suitable area for fruitful interchange. A British Nuclear Forum delegation, including NNC staff, therefore visited the USSR in October/November 1975 and there was also a return visit. The main objective of the visit was to see what could be learned from the large effort that was being made in the design and construction of the RBMK pressure tube type of reactor. At this time the UK was engaged on the design of the Steam Generating Heavy Water Reactor (SGHWR), also a direct cycle pressure tube reactor, as a commercial power producer and the superficial similarity between the two concepts suggested that closer collaboration could be beneficial. A report on these exchanges was produced by the NNC.

In the report, reservations were expressed on a number of safety related aspects of the RBMK design. Although the report was not passed to the Russians they were made aware of the reservations during meetings both in the USSR and in the UK and it is of interest that design improvements in the later RBMKs address some of these reservations. The safety reservations from the NNC report are summarised together with comments on the improvements incorporated in the Chernobyl design. Despite these improvements the Russians acknowledged in Vienna that the Chernobyl design had fundamental shortcomings and these are noted where appropriate.

The reservations highlighted in the NNC report were:

i the lack of a direct in-core spray emergency core cooling system;

ii the lack of full containment appropriate for a water cooled reactor;

iii the mechanical instability of the core;

iv insufficient protection against failure of a pressure tube;

v the reactor has a positive void coefficient;

vi insufficient shutdown margin;

vii the possibility of zonal instabilities and local criticality in the core;

viii no back-up to the control rods for reactivity shutdown;

ix the high temperature of the graphite core throughout the life of the reactor.

0.4 Post-Chernobyl Design Changes to RBMK Reactors

The Russians acknowledge that the RBMK design has shortcomings and they have described the design changes that they propose to make to remedy these. The accident occurred because the reactor could, and did, operate in an unstable regime and because it could not be shut down with sufficient rapidity. The design changes are intended to redress these deficiencies and are as follows:

Short term

1 The linkages between the control rods and graphite followers will be modified to ensure that a "positive scram" effect cannot occur.

2 The equivalent of 70-80 rods will be kept within the core.

3 Violation of the requirement to avoid operating below 700MW(th) will be prevented by additional shutdown protection.

Longer term

4 To mitigate the problem of the positive void coefficient fixed absorbers will be installed and

5 the fuel will be modified by increasing its enrichment from 2% to 2.4% to compensate for the reduction of reactivity caused by (4).

6 Another long term development is a fast acting shutdown system, options for which are being studied.

7 A number of additional indicators of the cavitation margin of the main circulating pumps are being installed,

8 and also a system for automatic calculation of reactivity with an emergency shutdown signal when the reactivity reserve falls below a specified level.

In more recent publications (cf at the International Conference on Performance and Safety), more details of the measures have been given. Following further analysis, the above mentioned measures will be implemented, but they will be tailored to the requirements of specific reactors. The principles underlying the general requirements have not changed.

These measures will have an adverse effect on the economics of RBMK reactors but as best we can judge they should prevent a Chernobyl-type accident recurring.

In addition the Russians intend to improve operator training and undertake additional R&D on the physics and thermohydraulics of the RBMK reactors.

0.5 The Accident at Chernobyl

The information on the accident at Chernobyl Unit 4 provided by the Russians at the IAEA Experts' Meeting in Vienna in late August, 1986, has greatly clarified the circumstances of the accident.

The reasons for the accident are now clear. Maloperation of the reactor resulted in the core containing water at just below the boiling point, but little steam at the start of a turbogenerator experiment. When the experiment began, half of the main coolant pumps slowed down and the flow reduction caused the water in the core to start boiling vigorously. The bubbles of steam that formed absorbed neutrons much less strongly than the water they displaced and the number of neutrons in the core started rising. Technically, such behaviour was a consequence of the "positive void coefficient of reactivity" of RBMK reactors. It was normally counteracted by the "fuel (Doppler) effect", which causes the number of neutrons in a reactor core to be reduced as the fuel temperature rises. At the start of the turbogenerator experiment at Chernobyl however, the conditions were such that the void effect was dominant and the "power coefficient" of the reactor was positive. Then as the number of neutrons and power of the reactor rose, more steam was produced and even fewer of the neutrons were absorbed, causing the power to escalate through a "positive feedback" phenomenon.

At this stage, the operator realised that the reactor power was rising and pressed the emergency shutdown button, but it was too late. The reduction of neutron absorption caused the "excess reactivity" of the core to rise to the level where the "chain reaction" could be sustained by "prompt neutrons" alone and the reactor became "prompt critical" before the slow acting shutdown system could become effective. The activation of the shutdown systems themselves may have introduced further reactivity via the positive scram effect. The power surge caused the fuel to heat-up, melt and disintegrate. Fragments of fuel were ejected into the surrounding water, causing steam explosions that ruptured fuel channels and led to the pile cap (ie upper biological shield and charge face) being blown off. It has been suggested that hydrogen formed by the chemical reaction of steam and zirconium was then able to mix with air in the reactor hall and explode, causing further damage to the reactor building.

In essence, the Chernobyl accident was a steam explosion (or rapid evolution of steam) triggered by a "prompt critical excursion".

The prompt critical excursion caused a rapid power surge. Some press reports have described this rapid power surge as a "nuclear explosion" but that is incorrect. A nuclear explosion in an atomic bomb depends upon the fission by fast neutrons of nuclei of uranium 235 or plutonium 239 ("fissile" materials) and on keeping the nuclei close together long enough for millions of fissions to occur very rapidly indeed. As a bomb uses pure fissile materials driven together by explosive forces, the power of a nuclear explosion can be enormously large. Such an event cannot happen in a thermal reactor however, because the fissile material is mixed with a much larger amount of non-fissile material (uranium 238) and also because material rapidly disrupts (as it did at Chernobyl) bringing the fission process to an end. In a power surge, the neutrons are slowed down by the moderator just as they are in normal operation. As the process depends upon fission by slow neutrons, the time between successive fissions is relatively long and the energy produced would disrupt the fuel and hence terminate the fission chain reaction long before the reactivity reached the very high levels achieved in an atomic bomb. The proportion of the uranium or plutonium undergoing fission and hence the energy released in a power surge would be far lower (volume for volume about 100,000,000 times less) than would be achieved in a bomb. The special feature of the Chernobyl reactor was that the power surge took place with positive feedback because of the inherent instability of the reactor concept. Therefore the power surge was exceptionally large. It is, however, incorrect to describe the Chernobyl accident as a nuclear explosion.

The Russians' analysis of the accident indicates that fuel degradation occurred while there was still water in the fuel channels, due to a heat transfer crisis at the surface of the fuel rods. The fuel then disintegrated and rapidly generated steam by mixing with the water present in the channels. The pressure consequently rose and caused the fuel channels to rupture. The massive pile cap was then blown off and core materials were ejected into the atmosphere. The emergency core cooling system capability was destroyed by the explosions.

Details of the manner of failure of the pressure tubes are possibly of secondary importance for an appreciation of the consequences of their failure. Nonetheless, to demonstrate the more robust response of other reactor systems were a similar fuel disruption event to occur, it is necessary to understand these details. Failure could have been caused by one of several types of fuel-coolant interaction, these being: solid fuel fragmentation followed by dispersal into water and rapid steam generation; a steam explosion; a steam spike.

The core inerting system was designed to relieve the pressure that would have resulted from a single channel rupture and was not able, therefore, to cope with the pressurisation that resulted when many channels burst. The reactor pile cap was blown off by steam pressure and all fuel channels were ruptured. The work done blowing off the pile cap is estimated to be in the range 0.2 to 2 GJ and could have been supplied by either fuel-coolant interactions or expansion of the coolant released from ruptured channels with no energy contribution from the fuel.

The containment was breached within seconds of accident initiation and burning core materials were ejected into the air above the reactor. The core was exposed to the atmosphere and the graphite moderator was set on fire. The Russians decided to drop materials onto it to mitigate the fission product releases. About 5000 Te of material was dropped, including boron carbide to ensure shutdown, dolomite (limestone), which would decompose, release carbon dioxide and starve the core of oxygen, lead to remove heat and clay and sand to filter and retain fission products escaping from the core. This material thermally insulated the core and after several days its temperature began to rise, reaching a peak of about 1900C around the 4-5 May and decreasing thereafter.

The ultimate decrease in temperature was attributed to improved core cooling resulting from circulating air and nitrogen injection. Chernobyl Unit 4 has now been entombed in concrete.

The accident at Chernobyl has revealed numerous deficiencies in the design of RBMK reactors when judged by the NII's safety assessment principles. These principles are not theoretical ideals that would be unobtainable in practice, but rather "are based on experience obtained so far on the operation of commercial plant in the United Kingdom and represent a level of protection against radiological consequences of normal operation and fault conditions that should in most circumstances prove to be reasonably practicable". The design deficiencies would have made RBMK reactors unacceptable in the UK and the Chernobyl accident does not, therefore, reveal any deficiencies in UK safety standards.

The accident has highlighted design deficiencies in six areas:

I positive void coefficient and positive power coefficient;

II inadequate shutdown system;

III scope for operator interference with safeguards;
IV inadequate instrumentation and alarms;
V propagation of damage on-site;
VI vulnerability of safeguards to fault conditions.

To summarise, it is seen that the accident happened because of:

i faults in the concept of the reactor (inherent safety not built-in);

ii faults in the engineering implementation of that concept (insufficient safeguard systems);

iii failure to understand the man/machine interface ("a colossal psychological mistake" in the words of Mr Legasov, the head of the Russian team at Vienna);

iv poor operator training.

These faults placed an intolerable burden of responsibility for the safety of the nuclear reactor upon the operators. They are symptomatic of the differences between the Russian and UK approaches to nuclear safety, which are highlighted throughout this Report by comments drawing attention to the many aspects of RBMK design that do not conform to specific safety assessment principles enunciated by HM Nuclear Installations Inspectorate. It is instructive to see how they would have been avoided by the application of UK nuclear safety principles.

i First, and most important, in UK reactors protection against reactivity faults is achieved by selecting design concepts with intrinsic characteristics which provide inherent protection. For example, UK gas cooled reactors do not have a void coefficient either positive or negative and the void coefficient in a PWR is either negative or so slightly positive that the power of a reactor as a whole cannot run away. However, the RBMK reactor has such a large positive void coefficient at low power that it can dominate the behaviour of the reactor and the reactor can be intrinsically unstable; that is, if the power increases spontaneously, then it increases still more and escalates to higher and higher values. This positive feedback phenomenon is not an inherently safe characteristic.

ii Secondly, in UK reactors, the natural defences provided by the intelligent choice of reactor concept are supplemented by engineering features to prevent, limit, terminate and mitigate faults. Thus, for example, it is physically impossible to withdraw the control rods rapidly and if the operator seriously mishandles them, the reactor automatically fails safe, ie shuts down. In other words, the engineering implementation of a UK design ensures that the reactor remains safe even if the operator tries to do the wrong thing. The RBMK reactor was the exact reverse of that. The Russian designers knew that their design was intrinsically unstable at low power, they knew that was potentially unsafe but they did not take any engineering steps to avoid that unsafe condition. They simply instructed the operators not to operate the reactor below 20% power and they relied upon the operators to follow that instruction faithfully.

Furthermore, in any reactor and especially one which was intrinsically capable of rapid positive feedback and power excursions, UK safety rules would insist on fast-acting control rods. The Russians also neglected to provide these. A reactor design that allowed a potential for a positive scram effect would be unacceptable in the UK.

iii Thirdly, for UK reactors, the man/machine interface is assessed very carefully. The information to be given to the operator in the control room is considered, the reactor is designed so that he does not have to make important decisions in a hurry and devices are fitted so that if the operator does make a mistake, the reactor fails safe. The Russians admitted that they had not previously realised the importance of these points.

iv Fourthly, in the UK, operator training is regarded as extremely important. The CEGB and SSEB insist that their operators be highly qualified and have regular refresher courses including training on simulators. The Russians admit that their training was inadequate.

v Finally, the commercial UK nuclear system is overseen by an independent nuclear inspectorate, which can at any time without hindrance or challenge close down any licensed reactor. The Russians do not appear to have had an effective independent inspection capability.

In conclusion, the Chernobyl accident was unique to the RBMK reactor design and there are few lessons for the United Kingdom to learn from it. Its main effect has been to reinforce and reiterate the importance and validity of existing UK safety standards.

0.6 Source Terms

The source terms describe the details of the radioactivity release in an accident, and other characteristics such as the timing and energy of release, which are needed in order to assess the environmental consequences. The Russian account of the source terms for the Chernobyl accident, presented at the IAEA meetings in Vienna, show that this was a severe accident in terms of the large fractions of the core inventory of potentially harmful radionuclides released. Soviet measurements lead to the conclusion that 100% of the noble gases, 10-20% of the volatile fission products iodine, caesium and tellurium, and 3-4% of all other radionuclides escaped to the environment over a 10 day period from the 26 April to the 5 May. In total, about 1.85 EBq (50MCi) of released activity was present in the environment on the 6 May . This extended release period is a notable feature of the Chernobyl source terms and contrasts strongly with the release periods of at most a few hours predicted in severe accident analyses of PWRs.

The Russian account divides the release period into four stages:

Stage 1: 26 April 01.24. The initiating in-core transient blew off the pile cap and ejected fragments of hot fuel, together with vapours of the volatile fission products, directly into the environment.

Stage 2: 26 April - 2 May. In-core fires promoted a high level of continuing activity release through the 26 and into the 27 April. Dumping of materials (sand, dolomite, clay etc) onto the core debris began on the 27 April, and led to a steady reduction in the activity release rate until the 2 May.

Stage 3: 3-5 May. The core temperature, driven by decay heat, rose during this period to levels where a steady increase in activity release occurred, especially due to iodine. A second peak in activity release occurred on the 5 May.

Stage 4: 6 May. A sharp fall in activity release to an insignificant level occurred, coinciding with the injection of high flows of nitrogen under the core debris.

Comparisons of the radionuclide compositions of Russian and Western European samples of the released activity with the core inventory predicted by the FISPIN code have been made. The composition of the active material near the reactor throughout the release period was close to that of whole fuel, enhanced by factors of 3-6 in the proportion of volatile fission products. Most of the fuel-based material settled out within a few tens of kilometres of the reactor site, and was probably particulate towards the upper end of the aerosol size range. Samples collected in Western Europe show a different composition, being composed almost entirely of the volatile fission products, with only a few percent of the less volatile fission products and fuel. This material may have been released as fine aerosols, probably in the initial explosion (Stage 1). It is estimated that at least 16% of the core inventory of iodine 131 and 11% of the inventory of caesium isotopes 134 and 137 crossed the Soviet border and were deposited in Eastern and Western Europe. Thus, the total release fractions for iodine and caesium exceeded 36% and 26% respectively. Apart from this omission, the Russian account of the source term agrees with the limited independent evidence available.

0.7 Environmental Consequences

A general picture of the dispersion of radioactive material from Chernobyl across Europe has been assembled, based on reported measurements conveyed through national and international bodies (IAEA, WHO and NEA). Wet deposition during periods of rainfall caused a marked patchiness in the environmental measurements from different countries.

Initially, activity was transported in a north-westerly direction from Chernobyl into Scandinavia. A few days later, however, more westerly trajectories were followed bringing the material across Europe. During this time, an area of high pressure was moving eastwards across the European continent. This resulted in fairly wide dispersal of the material and gave rise to a trajectory that brought radioactivity to the UK on the 2nd May. The contamination of the northern part of the UK was greater than that of the south, reflecting different rainfall rates during plume passage.

Estimates have been made of the dosimetric impact of the release on both Western and Eastern Europe, based on measured levels of activity in the environment. The main dosimetric pathways contributing to individual exposure are inhalation of activity from the plume, exposure to external radiation from deposited activity and ingestion of contaminated foodstuffs. The total collective dose, summed over all countries in both Eastern and Western Europe, is estimated to be approximately 1.8×10^5 man Sv. The average dose in the Eastern European countries is estimated to be approximately 3 to 4 times that in Western Europe. The average individual dose in the UK resulting from Chernobyl, integrated to 50 years, represents about a week or two at normal background dose rates or equivalently, to having a three-week holiday in Cornwall, where the background dose rate is slightly higher than in some other parts of the UK.

Chernobyl Bibliography

I National Reports

Australia

1 Australian Atomic Energy Commission - Research Establishment, Lucas Heights.
First Report of the Task Group on The Accident at the Chernobyl Nuclear Power Plant, 2 May 1986.

Canada

2 Atomic Energy of Canada Ltd
Chernobyl - A Canadian Technical Perspective
AECL - 9334.

Finland

3 Finnish Meteorological Institute - Interim Report
Dispersion of Radioactive Releases following the Chernobyl Nuclear Power Plant Accident. ISSN 0782-6079, 1986
Savolainen and others.

4 Finnish Centre for Radiation and Nuclear Safety
Interim Report on Fallout situation in Finland from April 26 to May 4 1986. ISSN 0781-2868.

5 Environmental Science and Technology
Radioactive Size Distribution of Ambient Aerosols in Helsinki, Finland, during May 1986 after the Chernobyl Accident: Preliminary Report. Kauppinen and others Vol 20 No 12 pp1257-1259. American Chemical Society.

France

6 Commissariat A L'Energie Atomique Institut de Protection Et De Surete Nucleaire.
The Accident at Chernobyl, Rapport IPSN No 2/86, Revision 2, June 1986.

7 Risley Translation 5335 (Rev 3 October 1986)
Report of the Minister for Industry on the Chernobyl Accident
Doc WG1-87/P20.

Germany

8 German Federal Parliament
Report on the Chernobyl Reactor Accident and its Consequences for the Federal Republic of Germany, GRS-S-39, June 1986.

Ireland

9 Nuclear Energy Board Dublin
Chernobyl, Its Effect on Ireland, March 1987
Cunningham and MacNeill.

Italy

10 ENEA Italy. The File on Chernobyl, Two Volumes, June 1986.

Netherlands

11 Netherlands Energy Research Foundation
ENC Report on the Chernobyl Accident prepared by the CCRX Commission for the Netherlands Government. Radioactive Contamination in the Netherlands resulting from the Chernobyl Nuclear Accident, October 1986.

12 Some Calculations on the Reactor at Chernobyl, W J Oosterkamp, 00651-RSR-87-3047.

Portugal

13 Radioactivity in Portugal following the Chernobyl Accident.

Spain

14 Technical Report UNESA Unidad Electria S.A
The Chernobyl Accident and the Spanish Nuclear Power Plants.
November 1986.

Sweden

15 Swedish Ministry of Industry
Report of the Expert Group for Nuclear Safety and the Environment After Chernobyl. Consequences for Energy Policy, Nuclear Safety, Radiological Protection and Environmental Protection.
Doc WG1-87/P21, July 1987.

UK

16 United Kingdom Atomic Energy Authority.
The Chernobyl Accident and its Consequences NOR 4200, Gittus and others. March 1987.

17 The Watt Committee on Energy
The Chernobyl Accident and its Implications for the United Kingdom:
To be published shortly, 1988.

18 The Central Electricity Generating Board.
Chernobyl. - J G Collier and L Myrddin Davies, October 1986.

19 National Nuclear Corporation
The Russian Graphite Moderated Channel Tube Reactor
Yellowlees and others NPC(R)1275, March 1976, renumbered as ACSNI(86)INF, 14 May 1986.

20 ApSimon Macdonald Wilson
An Initial Assessment of the Chernobyl 4 Reactor Accident Release Source. Journal. Society for Radiological Protection, Vol 6 (3) pp109-119 - 1986.

21 Simmonds J R
Chernobyl: Europe Calculates the Health Risk
New Scientist 114, 40-43 - 1987.

22 Fry F A and Britcher A
Doses from Chernobyl Radiocaesium
Lancet, July 18, 1987.

23 Clark and Smith
Wet and Dry Deposition of Chernobyl Releases
Nature Pub.

<u>USA</u>

24 US NRC NUREG-1250
Report on the Accident at the Chernobyl Nuclear Power Station, January 1987.

25 US NRC NUREG-1251
Implications of the Accident at Chernobyl for Safety Regulation of Commercial NPP's in the United States. Draft for Comment August 1987.

26 US Dept of Energy
Analyses of Chernobyl-4 Atomic Energy Station Accident Sequence. DOE/NE-0076, November 1986.

27 US Dept of Energy
Inter laboratory Task Group on Health and Environmental Aspects of the Soviet Nuclear Accident, DOE/ER-0332 IUC-41&48) 1987.

28 US Dept of Energy
Collective Radiation Doses from the Chernobyl Reactor Accident Calculated from an Atmospheric Dispersion and Deposition Model. Soldat and other, March 1987, DOE X87-019.

<u>USSR</u>

29 Information Compiled for the IAEA Information Meeting 25-29 August 1986, Vienna
The Accident at Chernobyl Nuclear Power Plant and its Consequences
Part I General Material
Part II, Annexes 1,3,4,5,6 and Part II, Annexes 2,7 all at draft.

30 Atomnaya Energiya, Vol 62, No 4 April 1987, pp219-226
Increasing the Safety of Nuclear Power Plants with RBMK Reactors.
Adamov and others.

31 IAEA Conference on Nuclear Power Performance and Safety, Vienna: 28 September to 2 October 1987.
The Chernobyl Nuclear Power Station Accident: One Year Afterward. IAEA-CN-48/63.65, Asmolov and others.

II <u>International Reports</u>

<u>International Atomic Energy Authority (IAEA)</u>

32 Summary Report on the Post-Accident Review Meeting on the Chernobyl Accident (IAEA) - Vienna.
Safety Series No 75-INSAG-1, 1986.

33 General Plant Description Chernobyl-4 (IAEA) Vienna
Jankowski and others. Doct. 6729n Rev 2 1986.

34 International Conference on Nuclear Power, Performance and Safety - Book of Abstracts - IAEA Vienna, 28 Sept to 2 Oct 1987, IAEA-CN-48.

<u>Nuclear Energy Agency - Organisation for Economic Co-operation and Development (NEA-OECD)</u>

35 The radiological impact of the Chernobyl Accident in the OECD Countries, Paris, Sept 1987.

36 Chernobyl and the Safety of Nuclear Reactors in OECD Countries, Paris 1987.

<u>World Health Organisation</u>

37 Chernobyl Reactor Accident, ICP/CEH 129
Report of a Consultation, 6 May 1986.

38 Analysis of the Radiological Consequences of the Accident at the Chernobyl NPP for the Population of the European Regions of the USSR, WHO Information Document, A40/INF.DOC./9, May 1987. This is a USSR Health Ministry Report.

NOTE

The authors of this Report have inserted Comments into the text where they have thought it appropriate. In the main, the Comments draw attention to the UK Nuclear Installations Inspectorate's Safety Assessment Principles (Ref 1). These Principles are intended to be used by the Inspectorate's staff as a basis for the assessment of designs for nuclear installations submitted to the Inspectorate as part of the review that must precede the licensing of nuclear installations in the UK. The designs and safety case of the Russian RBMK reactors were not available to us however, so we have had to infer aspects of these from information provided by the Russians in the past and from information about the accident at Chernobyl. Consequently, although every effort has been made in the Comments to judge the RBMK design fairly against the NII Safety Assessment Principles, accuracy at a detailed level cannot be guaranteed. Furthermore, it should be recognised that the NII's Safety Assessment Principles are not inviolable rules. They are guidelines for the NII's own staff that are expected to be met as far as is reasonably practicable. The NII's safety assessors judge the extent to which a design submitted to them conforms with their Safety Assessment Principles in their pre-licensing review. Nonetheless, they require justification of any adverse departure of a design from their Safety Assessment Principles.

In addition, the CEGB have developed their own Design Safety Criteria (Ref 2), which take into account the NII's Safety Assessment Principles, and some of the Comments refer to these. For particular projects, eg the PWR, these Criteria are amplified into CEGB Design Safety Guidelines for use by the Designer.

Both the NII Safety Assessment Principles and the CEGB Safety Criteria have been developed over the past ten years and are intended to provide guidance for the design and assessment of modern nuclear reactors. Chernobyl Unit 4 started construction in 1975/76 and it is therefore appropriate to review its design against these Principles and Criteria.

References

1 "Safety assessment principles for nuclear reactors", HM Nuclear Installations Inspectorate, HMSO, 1979.

2 "Design safety criteria for CEGB nuclear power stations", CEGB Health and Safety Department, HS/R167/81 (Revised), 1982.

SECTION 1: USSR POWER REACTOR PROGRAMME

Nuclear power is essential to the economy of the USSR. At the end of 1985 their total nuclear generating capacity was 26 GWe which was 15% of the total electricity production. The Soviet's plan to double their nuclear power generating capacity by 1990 and to increase it by a factor of five by the year 2000.

The USSR employs two main types of power reactor: the RBMK and the VVER. The RBMK with which this report is concerned has a graphite moderator, pierced by vertical holes through which zirconium alloy pressure tubes pass, each containing a fuel-stringer. Light water is pumped into the lower ends of the pressure tubes and boils as it passes over the fuel. Steam mixed with water emerges from the tops of the pressure tubes. Thus the RBMK is a boiling water, pressure tube, graphite-moderated reactor.

Table 1 lists the large RBMK reactors that were in service or under construction at the end of 1985. They have been developed from a small unit designed over thirty years ago. The fourth 950 MWe unit at Kursk and the second 1450 MWe unit at Ignalina have entered service during 1986. At Chernobyl units 1 and 2 are back producing power after decontamination of the site and buildings, unit 3 is still being decontaminated prior to re-commissioning, unit 4 has been entombed and units 5 and 6 which were under construction may not now be completed. Tables 2 and 3 give information for the Soviet's VVER's. Fig 1 shows the location of the Soviet reactors and reactors in European countries.

The analysis to which nuclear power has been subjected, following the Chernobyl accident, has not led the Soviets to any change in their position in relation to the development of nuclear power in the USSR. The Soviet's plans to introduce further nuclear power plants has not changed significantly. The remaining RBMK reactors under construction, will be completed (with the possible exception of Chernobyl units 5 and 6), but the Soviet's have said that their nuclear programme will then concentrate on VVER units.

TABLE 1

Large RBMK Units in Service and Under Construction in USSR

Status at 31.12.85	Station	Unit Output MWe (net)	No of Units	Commercial Operation (Commencing dates)
In Service	Leningrad	950	4	1974-1981
	Kursk	950	3	1976-1983
	Chernobyl	950	4	1978-1984
	Smolensk	950	2	1983-1985
	Ignalina	1450	1	1984-
Under Construction	Kursk	950	1	1986-
	Ignalina	1450	1	1986-
	Chernobyl	950	2	1986-1989
	Smolensk	950	2	1988-1989
	Kostroma	1450	2	1988-1989

TABLE 2

VVER Units in Service in USSR

Status at 31.12.85	Station	Unit Output MWe (net)	No of Units	Commercial Operation (Commencing dates)
In Service	Novo Voronezh	265	1	1964-
		338	1	1970-
		410	2	1972-1973
		953	1	1981-
	Kola	440	4	1973/75-1982/1984
	Armenia	370	2	1976-1980
	Rovno	420	2	1981-1982
	Nikolaiev	953	2	1984-1985
	Kalinin	953	2	1984-1985
	Bala Kovo	953	1	1985-
	Zaporozhe	953	1	1985-

TABLE 3
VVER Units Under Construction in USSR

Status at 31.12.85	Station	Unit Output MWe (net)	No of Units	Commercial Operation (Commencing dates)
Under Construction	Zaporozhe	953	5	1986-1991
	Khmelnitski	953	4	1986-1990
	Nikolaiev	953	2	1987-1989
	Aktash	953	2	1987-
	Tatar	953	1	1987-
	Volgodonsk	953	4	1987-1990
	Rovno	953	2	1988-1990
	Bashkir	953	2	1988-1989
	Odessa	953	2	1988-1990
	Balakovo	953	2	1989-1990
	Nizhinekamsk	953	1	1989-

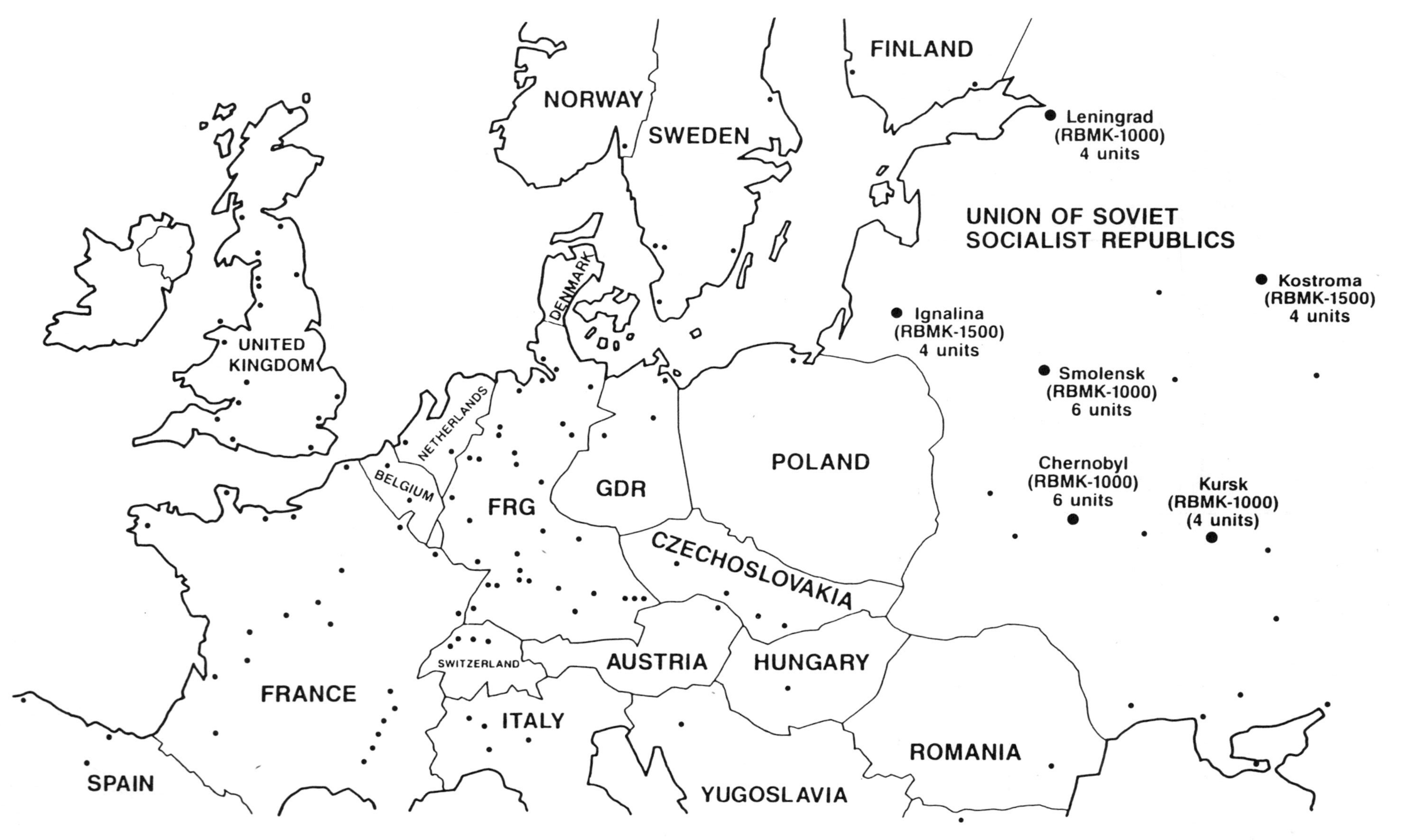

FIG. 1 LOCATIONS OF RBMK STATIONS IN THE USSR AND OF OTHER NUCLEAR POWER SITES IN NORTHERN AND MIDDLE EUROPE

SECTION 2: DESCRIPTION OF THE RBMK REACTORS

The Chernobyl plant has RBMK boiling water, pressure tube, graphite-moderated reactors. The combination of a pressure tube coolant circuit with a graphite moderator in a commercial nuclear power station is a hybrid, unique to the Soviet Union. Its parentage can be traced to the early reactors built to produce military plutonium.

The chief design features of the RBMK reactors are:

a Vertical pressure tubes with onload refuelling.

b Fuel assemblies embodying 18-pin clusters, each pin comprised of uranium dioxide fuel pellets clad in a zirconium alloy tube.

c A graphite moderator and reflector.

d Boiling water coolant, the steam going to the turbines.

In this Section salient features of the RBMK design are described with comments regarding their adequacy against UK safety practice with particular reference to NII Safety Assessment Principles (Ref 2). Relevant parameters are summarised in Tables 1 and 2. Unless otherwise stated the information is from Ref 1.

2.1 Overview of the RBMK

The main features of the RBMK are shown in Fig 1. At the centre of the unit is the reactor core with its supporting structures and biological shielding. The reactor coolant circuit, consisting of a complex array of pipework and valves, supplies water to the fuel channels and transports steam/water mixture to the steam drums. Above the reactor is the reactor hall, containing the fuelling machine. A containment building partially surrounds the reactor and primary circuit.

An inherent feature of the design is that the generation of steam in the fuel channels makes a positive, destabilising, contribution to the reactivity of the core. In normal operation this effect is countered by the negative Dopper fuel temperature coefficient. These characteristics and the limits of operation required to ensure a stable reactor operating regime are discussed later in this Section.

2.2 Reactor and Supporting Structures

The reactor and its supporting structures are shown in Fig 2. The reactor core is cylindrical in shape, with a diameter of 11.8m and a height of 7m excluding reflectors. It consists of graphite blocks assembled into columns, with vertical, cylindrical openings which house a total of 1661 fuel channels, 211 control channels and 12 axial detector strings. The maximum temperature of the graphite is reported to be about 700°C.

COMMENT [**This graphite temperature is somewhat higher than those of which the UK has experience. It is not known whether the USSR have qualified the graphite blocks for the full range of temperature, humidity and irradiation conditions over the design life of the reactor. UK experience suggests that the RBMK conditions are very onerous for the graphite.**]

The whole of the reactor is located in a leaktight cavity (formed by the upper and lower shields and a cylindrical steel shell), designed to withstand an internal pressure of 0.08 MPa (11.6 psi). The cavity is filled with helium/nitrogen mixture (85-90% He, 15-10% N_2) which inhibits oxidation of the graphite and improves heat transfer from the graphite to the cooling channels. The He/N_2 mixture is circulated through a clean-up system where it is monitored for temperature and moisture content.

COMMENT [**The relatively high graphite temperature adds to the importance of excluding air from the reactor cavity.**]

Outside the reactor cavity is a lateral shield which takes the form of a cylindrical reservoir (annular in cross-section) 19m outside diameter and 16.6m inside diameter. The tank is fabricated from low alloy steel, and is divided internally into 16 vertical leak-tight compartments filled with water. The tank is supported from below by the building structure.

The upper biological shield is a fabricated steel structure, 17m in diameter and 3m in height. It is perforated by standpipes which accommodate the fuel and control channel assemblies. The space between the standpipes is filled with serpentinite concrete. The upper shield supports the loaded fuel channels, the floor of the reactor hall above, and the reactor coolant outlet piping. This entire structure is in turn supported on roller-type supports by the lateral shield tank.

The lower shield is similar in design to the upper structure, but 14.5m in diameter and 2m high. The structure supports the graphite stack and the reactor coolant inlet piping, and is perforated by steel tubes which accommodate the lower ends of the fuel and control channels. The lower shield is supported from below, by a fabricated steel structure, which transmits the weight of the graphite stack to the building.

Above the upper shield is a gallery containing the upper ends of the standpipes and the reactor outlet (or riser) pipes. Above this are slabs which form biological shielding for the reactor hall, and act as heat insulation. The slabs rest on the standpipes and can be removed to permit access to the standpipe for refuelling.

The fuel channel, illustrated in Fig 3A consists primarily of zirconium alloy (Zr, 2.5% Nb) tube, 88mm external diameter with a wall thickness of 4mm. Steel/zirconium alloy transition pieces facilitate the attachment of upper and lower steel end-pieces. The upper end piece is rigidly attached to the steel standpipe which penetrates and is sealed to the upper shield structure. The lower end piece is sealed to the lower shield structure through a bellows unit which compensates for the difference in thermal expansion between the fuel channel and the reactor cavity structure.

The fuel assembly, Fig 3A, is mounted in and suspended from the top of the channel tube by a ball-type plug that seals the top of the channel tube. Two fuel sub-assemblies, each 3.5m long are held together by a separate support rod and suspended from the plug. Each sub-assembly comprises 18 fuel pins spaced by 10 stainless steel cellular spacer lattices (see Table 2).

The fuel is replaced on load. The charge machine carries out the sealing and unsealing operation remotely.

The control rod channel heads are designed to accept the attachment of actuators and also the inlet cooling water supply from a cooling system that is separate from the main fuel channel cooling.

The fuel channels are the principal means of heat removal from the graphite blocks. Graphite "piston rings" fit around the pressure tube, alternately in contact with the pressure tube and the graphite block, as illustrated in Fig 3B. The cooling of the graphite depends upon the heat conduction path through the piston ring arrangement.

COMMENT

The gap between the fuel channel and the graphite is small, so that failure of the pressure tube would result in significant pressure loads on the graphite blocks. This would tend to lift the bricks and would be likely to damage them, particularly after irradiation. Consequently the graphite stack appears to be vulnerable to pressure tube failure, and there is a likelihood of propagation of failure to other tubes. Taken together, the relatively low design-pressure of the reactor cavity, the lack of containment above the reactor and the pressure tube/graphite moderator design appear vulnerable and could lead to a significant release of radioactivity in an accident that included pressure tube failure.

The risk of failure propagation and breach of the reactor cavity are contrary to NII Principles 91 and 152 respectively.

2.3 Primary Coolant Circuit

The primary circuit (illustrated in Figs 4 and 5) supplies water to the bottom of the fuel channels and removes the steam/water mixture from the top. It consists of two similar loops, which function in parallel and independently; each loop removes heat from half of the reactor's fuel assemblies.

In each loop, water from the suction header passes through four pipes to the main circulating pumps. Under normal operating conditions at high power, three of the four main circulating pumps are in operation with one held in reserve. Water leaves the main circulating pumps at a temperature of 270°C and a pressure of 8.4 MPa (1218 psi) and flows through pressure pipes, in each of which are installed in sequence: a non-return valve, an isolation valve and a throttle valve. Coolant then flows into the main circulating pump pressure header, from where it passes into the 22 distributing headers, which have non-return valves at their inlets. An individual line leads into each of the 1661 fuel channels. The flow rate through each channel is controlled by means of isolating and regulating valves. As it passes through the fuel channels, the water surrounding the fuel elements is heated to saturation temperature, partially evaporates (14.5% by weight on average) and the steam/water mixture at a temperature of 284.5°C and a pressure of 7 MPa (1015 psi) flows through the individual riser pipes into the steam drums where it is separated into steam and water. The water which has been separated out is mixed at the drum outlets with feed water, and flows through 12 downpipes (from each drum) into the suction header.

The temperature of the water flowing into the suction header depends on the rate of steam production of the reactor. When this decreases, the temperature increases somewhat because of the changing ratio of water from the drum separators, at a temperature of 284°C, and feed water, at a temperature of 165°C. Consequently, when the reactor power is below 500 MW ie during start-up or shut-down operations, the flow rate through the primary circuit is controlled by using throttle-type control valves and reducing the number of pumps operating to reduce the flow from the normal rate of 8000 m^3/hr per pump, to the range 6000-7000 m^3/hr. This is necessary to ensure that the temperature at the main circulating pump inlet is low enough to ensure that cavitation does not occur in the pumps and to maintain steam production in the core.

COMMENT

The pressure tube concept employed in the RBMK results in a much more complex arrangement than has become customary in the UK. Particular difficulties are:

a the potential for failure of the complex array of pipework;

b the potential for propagation of such failures from one pipe to another, in view of the close proximity of the pipes to one another and the difficulty of providing restraint of, or segregation between pipes;

c the complexity of the Emergency Core Cooling Supply (ECCS) needed to cater for breaks in any part of the circuit.

The RBMK pressure circuit is known to have little by way of restraint or segregation. The ECCS is designed only for single failures of specified sizes of pipe.

The vulnerability of the primary circuit is contrary to NII Principle 91. It would also make compliance with Principle 109, on adequacy of cooling at all times, difficult to demonstrate. Principle 46 for inspection would be difficult to meet.

2.4 Emergency Core Cooling System (ECCS)

The ECCS is designed to maintain core cooling in the event of pipe-failures in the main coolant system and in some "intact circuit" faults in which the supply of water to the fuel channels is interrupted for other reasons, [for example: loss of feedwater or loss of power supplies]. Actuation of the ECCS is normally initiated automatically, although it is known that the operators blocked ECCS operation prior to the accident at Chernobyl. Because the reactor coolant system is arranged in two separate, independent loops, the ECCS must be arranged to meet the differing functional requirements associated with the breached and intact halves of the reactor. The Chernobyl ECCS consists of three sub-systems (see Fig 5) all connected to the distribution headers of the coolant system. One sub-system provides flow to the damaged half only, during the early stages of a breach, during which time, cooling of the undamaged half is via the normal RCS route. The other two ECCS sub-systems provide long term cooling for the reactor, one sub-system for each half.

Short term cooling of the breached half is provided by three trains of ECCS equipment, comprising two accumulator trains and one pumped train. Each accumulator train consists of six accumulator tanks, containing water with a nitrogen gas blanket, maintained at 10 MPa (1450 psi), and is capable of delivering 50% of the maximum flow requirement to the damaged half of the reactor, for not less than 100 seconds after the initiation of the breach.

Fast acting valves are used to initiate ECCS flow within 3.5 seconds of the break in normal water supply. The third train utilises the feed pumps, which are electrically driven and which draw water from the deaerators. The pumps are realigned to feed directly to the ECCS headers. This route is capable of delivering 50% of the maximum flow requirement to the damaged half of the reactor.

Long term cooling of the damaged half of the reactor is provided by three trains of ECCS equipment. Each train consists of two pumps connected in parallel and is capable of delivering 50% of the maximum flow. The pumps draw water from the pressure suppression pools beneath the reactor, the water being cooled by service water in heat exchangers in the pump suction lines.

Long term cooling of the intact half of the reactor is provided by three trains of ECCS equipment. Each train consists of a single pump drawing water from a condensate storage tank, and is capable of delivering 50% of the required flow.

The ECCS is triggered by the coincidence of a high pressure signal from compartments containing the main coolant pipework and either a low level signal from the steam drums or a signal indicating a decrease in circuit pressure drop. The latter identify the damaged half of the reactor.

It is a requirement that the ECCS must fulfil its function in the event of a loss of normal power supplies coincident with the "design basis accident". The normal power supplies for pumps and valves in the ECCS are derived from the grid, but in the event that the grid is lost then power can be obtained from the running turbo-generator(s). During turbine run-down, the turbo-generator continues to support the feedpumps which contribute to short term ECCS duty for 45-50 seconds. Subsequently, diesel generators are brought in, to power the pumps in those sub-systems which contribute to longer term cooling. ECCS valves which cannot accept interruption in supplies are supported by batteries.

The emergency systems provide 150% redundancy. The design is claimed to take account of unproductive loss of coolant through the breach, and a single active or passive failure within or outside the ECCS.

COMMENT **Unavailability of part of the system due to maintenance does not appear to be taken into account as would be required in the UK. Whilst a detailed reliability assessment has not been made, the complexity of the system and its dependence on valve alignment would probably result in lower reliability than would be required in the UK, Principles 37 and 121 and Ref 4. The ECCS does not cater for stagnation accidents although such occurrences are deemed to be possible in pressure tube reactors: Principle 109 would therefore not be met.**

2.5 Containment Systems

The Chernobyl building structure is designed to contain and confine the release of radioactivity following failures in certain parts of the reactor coolant circuit. The containment structure and associated systems mitigate the effects of pressure-part failures in the lower parts of the downcomer, the pumps, the pressure headers, distribution headers and inlet pipework. However, failures in the upper parts of the fuel channel, the riser pipes, the steam separators themselves and the upper portions of the downcomers are not catered for. The boundaries of the containment structures together with the primary circuit are shown in Figs 5 and 6.

The principal features of the containment system are shown and numbered in Fig 6. These are

a leaktight compartments enclosing the pumps (labelled "1") designed for an overpressure of 0.45 MPa (65 psi).

b connected to the leaktight compartments: a steam distribution corridor (4) and a two-storey pressure suppression pool (5), partially filled with water. The partitions are perforated by non-return valves and venting channels.

c compartments (2) enclosing the distribution headers and reactor inlet piping which will tolerate an overpressure of 0.08 MPa (11.6 psi) and which are vented to the steam distribution corridor via non-return valves.

d the reactor cavity (3) which will tolerate an overpressure of 0.08 MPa (11.6 psi) is vented to the pressure suppression pool.

e the riser pipe gallery (6) and the steam drum compartment (7), which are not designed to withstand overpressure are normally maintained at a pressure slightly below that in the reactor hall.

Following pressure circuit failures in the pump compartments or the group header compartments, steam/water and air are vented to the suppression pools via non-return valves and venting channels as shown in Fig 7. The results of calculations presented by the Soviets show that the compartment pressures remain below the design values for the cases of failure of a pump pressure header, a distribution header or a single fuel channel within the reactor cavity. A separate venting route is provided for the reactor cavity, as illustrated in Fig 8.

There is provision for heat removal from the containment compartments and from the pressure suppression pools. Surface condensers in the steam corridor remove heat only during accident conditions. A sprinkler system operates both during normal operation and during accidents. Water is drawn from the suppression pool and is cooled in a service water heat exchanger. The water is then directed to the air space above the suppression pool, where it mixes and cools the air, and to jet coolers in the upper (hottest) part of the containment compartments. The jet coolers entrain air, thereby cooling the environment and removing radioactive aerosols and steam. The water is collected and returned to the suppression pool.

Provision is made to maintain the concentration of hydrogen below 0.2% (by volume). Hydrogen is present in the containment volume during normal operation owing to coolant leakage (assumed to be at a rate of up to 2 t/hr). During accident conditions hydrogen may also arise from a zirconium-water reaction. To cater for this, air is drawn from the containment volume at the rate of 800 m^3/hr during normal operation, and is discharged to atmosphere via filtration plant. The purge is automatically discontinued immediately following a failure of the coolant circuit and is then reinstated manually after a period of 2-3 hours, as hydrogen accumulates.

COMMENT [**NII Principle 152 requires provision of a containment around the reactor and its primary coolant circuit unless it can be shown that adequate protection has been achieved by some other means. The RBMK reactor has partial containment of the coolant circuit and this would probably fail to meet UK requirements given the vulnerability of the primary circuit (see Section 2.3 above).**]

2.6 Power Supply Systems

The Chernobyl power systems for Number 4 reactor are illustrated in Figs 9 and 10. The station has two main generators connected to the 750 kV grid via a single generator-transformer. Two generator switches are installed in series with each main generator and a connection for a unit transformer is made between each pair of generator switches. The plant also has a 330kV connection to the grid via a station transformer, although this route is normally isolated from the main electrical system. These power supplies feed the main distribution boards (to which loads such as reactor coolant pumps and feed pumps, are connected) and the essential distribution boards (to which essential drives, such as ECCS pumps or containment spray pumps, are connected).

Normal power supplies for house loads are obtained via the unit transformer. The following power sources are also available in various operating modes:

a following a reactor or turbine trip, supplies can be obtained from the 750kV grid via the generator transformer.

b following loss of the 750kV grid or generator transformer supplies can be obtained from the 330kV grid via the station transformer. This route is normally isolated and it is not clear under what circumstances it would be connected, nor how long the connection would take.

c following faults initiated by reactor coolant circuit failure with coincident loss of normal power supplies and turbine trip, the running down turbogenerators can be used to support electrical system loads for a limited time. This mode of operation is used to support the feed pump in its ECCS mode for a period of 45-50 seconds.

d following loss of all off-site supplies, essential loads can be driven from diesel generators.

Following a loss of off-site power, any running essential loads are tripped, the diesel generators are started automatically and connected to the electrical system within 15 seconds. Essential drives required by the incident are sequence started from the available power source (grid or diesel generator). Auxiliaries which cannot accept the interruption of supplies are supported by batteries. The three groups of essential safety equipment and their controls have independent power supplies, as illustrated in Figure 9.

COMMENT

The Essential Electrical System plays a major role in ensuring core cooling following faults such as breaches of the pressure circuit. Unavailability of one train of Essential Electrical Power owing to maintenance, in combination with a single failure in a second train, would result in inadequate core cooling capacity from the one remaining 50% train. The system would not meet NII Principles 37, 112, 121 or the CEGB criteria, Ref 4.

2.7 Neutron Flux Monitoring

The neutron flux detectors required for monitoring, control and protection are located in-core and ex-core as described below.

4 fission pulse detectors (KNT 31) are lowered into the reflector for start up. They are withdrawn when the next measurement range is reached. They cover the range 8×10^{-12} to 3×10^{-7} of full power.

4 sensitive ion chambers (KNK56) located in the shield tank give period protection during start up from 3×10^{-8} to 5×10^{-2} of full power.

4 ion chambers (KNK56) located in the shield tank supply the signals to control the four low power automatic rods (system AR2) in the range 6×10^{-3} to 1.5×10^{-1} of full power.

8 ion chambers (KNK53M) located in the shield tank supply the signals to control the two sets of four automatic control rods (systems AR1 and AR2) in the range 5×10^{-2} to 1.2 of full power. The signals are also used for power overshoot protection.

4 ion chambers (KNK53M) located in the shield tank supply signals for power monitoring and period protection in the range 10^{-6} to 1.2 of full power.

2 ion chambers (KNK53M) located in the shield tank supply signals for linear power reading in the range 10^{-2} to 1.2 of full power.

2 ion chambers located in the shield tank supply signals for linear power recording with a switchable range from 8 x 10^{-8} to 1.0 of full power.

24 fission ion chambers (KTV17) located in the core in pairs close to the 12 local automatic control rods supply the signals for control in the 12 zones and for insertion of the 24 local emergency protection rods.

130 flux monitors located at the positions shown in Fig 11, supply signals used in the calculation of the spatial power distribution.

12 flux monitors each with seven axial sections measure the axial flux shape at the positions shown in Fig 11 and supply signals used in the calculations of the spatial power distribution.

COMMENT [**There is considerable sharing of neutron flux instrumentation between control and protection functions. It is therefore unclear whether NII Principle 120 would be satisfied. That Principle requires that "the protective function should not be jeopardized by other functions".**]

2.8 Reactivity Control

The 211 control rods have the duties described below:

12 local automatic control rods for automatic control of the radial power shape. Withdrawal is inhibited if one of two flux detectors gives a high signal.

12 automatic control rods operating in three groups of four for overall power control. The rods in each group move in unison. One group (AR3) is for low power operation. The two remaining groups (AR1) and (AR2) are for normal power, one in operation, one is standby.

24 local emergency protection rods, two rods in each of the twelve auto control zones. They are inserted in response to a high local flux signal from the two detectors of the local auto control system until one signal clears.

24 part length rods inserted from the bottom of the reactor for manual control of axial flux.

139 manual control rods for power trimming. (24 of these, uniformly distributed are withdrawn prior to start up for emergency protection).

On a reactor trip all rods except the part length rods are inserted at 0.4 m/sec.

COMMENT **The effectiveness of the rod insertion is dependent on a sufficient number of rods being partly inserted into the core which would not meet Principle 143 to have an adequate shutdown capability at all times. A single system would also fail to meet Principles 121 and 122 which require diversity for systems that demand high reliability. Considering the characteristics of the reactor a fast acting system is necessary.**

2.9 Reactor Protection

The RBMK design specifically attempts to minimise plant outage from trips. For this reason there are 5 levels of reactor protection:

PROTECTION LEVEL	ACTION
5	Trip all rods (except the short bottom entry type), proceed to shutdown.
5*	Trip all rods (except the short bottom entry type), until the trip signal has cleared.
3	Rapid ramp down to 20% power
2	Rapid ramp down to 50% power
1	Rapid ramp down to 60% power

Note that the trip-action motors the rods into the core at 0.4 metres/second, giving a time to rod-bottom of the order of 18 seconds for any fully withdrawn rods.

Protection level 5 is activated by:

A power overshoot of 10% of nominal full power

A reduction in the reactor period to 10 seconds

Drum separator level high or low

Low feedwater flow

Excess pressure in leaktight compartments; drum separators, reactor cavity or lower water lines.

Low control-rod coolant-reserve or coolant-flow

Trip of two turbogenerators

Trip of the only operating turbogenerator

Trip of 3 out of 4 operating, main circulating pumps in either pump room.

Voltage loss in the plant auxiliary-power supply-system

Failure to respond to protection level 3,2 or 1 demands

Manual trip.

Protection level 5* is activated by an emergency power overshoot, generating a partial trip, stopping the rods when the overshoot signal clears. The distinction between a power overshoot activating level 5, and an emergency power overshoot activating level 5*, is not clear. With such heavy reliance on the operator to maintain a stable configuration and to carry out so many manual rod adjustments, it is believed that level 5* is the normally-activated level in response to power overshoot, with full trip by level 5 as the more extreme backup.

Protection level 3 is activated by:

Load rejection by both turbogenerators
Load rejection by the only operating turbogenerator

Protection level 2 is activated by:

Outage of one of two turbogenerators
Load rejection by one of two turbogenerators

Protection level 1 is activated by:

Loss of 1 of 3 operating main circulation pumps in either pump room.

Reduction of water flow in the primary circuit.

Reduction of feedwater flow.

Reduction of level in the drum separators.

Actuation of the group closure key for the coolant circuit throttle regulating valves.

COMMENT **Prior to the Chernobyl accident a number of trip parameters were blocked showing that the protection system would not meet NII Principles 38 and 131 on unauthorised access. Principle 38, in particular, says "Unauthorised access to and interference with safety-related structures, systems and components should be prevented by suitable measures". The total protection system is therefore not adequate (Principle 107). Detailed information and analysis would be required to assess compliance with Principles 121 and 122 which say that to cater for faults or maloperations the protection system should embody diversity and redundancy.**

2.10 Reactor Operation

The detailed safety design philosophy of the RBMK reactors is not available. However, the physics characteristics outlined in the Appendix illustrate why the normal operating regime must be constrained by the requirement that the positive void coefficient is limited in magnitude, leading to a negative value of the power coefficient.

These requirements are affected by power level and a combination of control rod worth and insertion. Hence:

a The worth of control rods inserted during normal power operations should be sufficient to restrict the reactor void coefficient to a level consistent with a negative power coefficient. An on-line computation of the 'operating reactivity margin' is available to the operator on demand. If he sees that it lies outside a limit (stated to be at least 30 equivalent rods inserted in the core) then he is supposed to trip the reactor. The operating reactivity margin is also quoted as "1% worth of inserted rods". The enforcement of this limit is the responsibility of the operator and no specific design provisions are made to ensure compliance.

b Sustained power operation is permitted only above a lower limit of 20% of full power (Ref 3) and operation below this level is subject to special restrictions. Again enforcement of this limit is in the hands of the operator and again there are no specific design provisions to ensure compliance.

COMMENT **In the UK, the intention is that the safety case should be established for the entire range of reactor characteristics, by modifying the design and/or the fuel loading so that only acceptable characteristics are built into the reactor. As a minimum requirement, it would be necessary to ensure, by the provision of interlocks or protection, that the operator could not**

inadvertently alter the reactor characteristics so that they seriously deviated from the assumptions presented in the safety analysis. The RBMK designers did not design or protect against the possibility of unsafe maloperations. They left the operator to trip the reactor manually if there was a risk that the power coefficient would become positive.

NII Principle 57 requires that the design characteristics should be stable and have no sudden change outside the specified operating limits. In particular Principle 68 addresses coolant voiding. Principle 43 requires that the design should prevent any operating mode exceeding safe limits. Principle 67 requires that movement of core components that could increase reactivity should be controlled by design. The RBMK design would not meet any of these requirements.

To trip the reactor, the control rods are driven into the core electrically at 0.4 metres per second, taking some 18 seconds to traverse the core length. The restriction in (a) above is stated to ensure the required initial trip worth of β per second, where β is the reactivity equivalent of the delayed neutron fraction, about 0.55% at equilibrium fuel cycle.

COMMENT **The RBMK protection system is unusual since it includes a level which allows entry of the rods to stop if the initial trip condition clears.**

It seems probable that the emphasis on maintaining reactor availability has influenced the operator's attitude to the importance of trip functions.

Readings from the 130 radial and 12 axial in-core detectors are processed in a centralised monitoring system, known as SKALA, which combines the detector readings with reactor calculations provided periodically by links to an external computer to derive the integrated power of each fuel assembly and the associated axial power profile. These power distributions are combined with measured assembly flow rates and inlet conditions to calculate the margin to critical heat flux in each assembly, to ensure no departure from nucleate boiling, which would lead to fuel overheating.

Unacceptably low margins to critical heat flux in any assembly are displayed to the operator on a reactor mimic plan every 5 to 10 minutes, together with a recommendation on flow adjustment to the assembly, while the automatic control system initiates a local or reactor power setback at zero margin.

The flow to each assembly can be adjusted by the operator and, in equilibrium fuel cycle, an adjustment is made at least twice in the life of each of the 1661 fuel assemblies, to optimise the reactor output.

Stability of the reactor power and power distribution in normal operation is controlled automatically by two discrete groups of rods:

1 the automatic mean power regulating group. Either of two groups of four control the reactor power in response to signals from detectors in the lateral water shield outside the core

2 the local automatic control group of rods divided into 12 single rod zones, controlling instabilities in the radial azimuthal power distribution, in response to signals from local detectors. This control function is backed up by a local emergency protection system, which either freezes the automatic rod withdrawal or inserts local emergency protection rods.

The operator is expected to trim the radial azimuthal power distribution by movement of any of 139 manual rods to optimise power output and also to eliminate the first harmonic in any axial instability by movement of manual rods into the top of the core or 21 shortened absorber rods into the bottom of the core, as appropriate.

The SKALA monitoring system computes the total reactivity worth of all control rods inserted into the core, by axial statistical weighting using the axial profiles from the axial detector strings. The result is compared with the operative reactivity margin of 1% and an indication is provided to the operator of any shortfall so that corrective action can be taken. The same result is used to assess the initial reactivity worth available for reactor trip, for comparison with the required minimum of 0.55% per second.

The role of the operator in high power operation is therefore to:

i adjust coolant flow through individual fuel assemblies.

ii control axial instabilities by movement of two types of absorber.

iii manually trim the radial-azimuthal power distribution to optimise power output, consistent with the automatic control by the local automatic rods.

iv operate such that the operative reactivity margin is maintained and therefore the void and power coefficient are within the safety design limits.

v in conjunction with (iv), ensure effective controller action by holding the automatic control rods within prescribed insertion limits.

vi in conjunction with (iv), maintain the initial trip requirement of at least 0.55% reactivity worth per second.

COMMENT **From the above it can be seen that the design places an unreasonable onus on the operator to ensure that the plant is within safe operating limits. It also allows access to important protection systems with the possibility of invalidation of the safety function. Neither of these aspects would be acceptable in the UK.**

2.11 References

1 "The accident at the Chernobyl nuclear power plant and its consequences", information compiled for the IAEA Experts Meeting, 25-29 August 1986, Vienna, by the USSR State Committee on the Utilization of Atomic Energy.

2 "Safety assessment principles for nuclear reactors", HM Nuclear Installations Inspectorate, HMSO, 1979.

3 "INSAG summary report on the post-accident review meeting on the Chernobyl accident", Vienna 30 August - 5 September 1986.

4 "Design safety criteria for CEGB nuclear power stations", CEGB Health and Safety Department, HS/R167/81 (Revised), 1982.

TABLE 1

General Specification

Thermal power, MW	3200
Electrical power (at generating terminals, MW)	1000
Active core diameter, m	11.8
Active core height, m	7
Lattice pitch, mm	250 x 250
Number of fuel channels	1661
Number of control rod channels	211
Constant uranium dioxide charge, t	204
Uranium, enrichment, %	2.0
Maximum design channel power, kW	3250
Power of most highly rated channel, kW	2700
Coolant flow in maximum power channel, t/hr	28
Coolant flow t/hour	37.5×10^3
Maximum steam content wt %	20.1
Mean bulk steam content wt %	14.5
Saturated steam temperature, deg C	284
Coolant temperature at fuel channel inlet, deg C	270
Saturated steam pressure in drum separators kgf/cm^2	70
Feedwater temperature, deg C	165
Maximum permissible graphite temperature, deg C	750
Burn-up MWD/kg uranium	20

TABLE 2

Characteristics of RBMK-1000 Fuel Sub-Assembly and Fuel Element

Distribution of fuel elements in fuel sub-assembly	2 rings of 6 and 12
Spacer grid	stainless steel cellular type
Supporting central rod	Zr alloy with 2.5% Nb
Weight of uranium dioxide (mean)	3.59 kg
Filler gas	Helium at 1 atm
Fuel element cladding	Zr alloy with 1% Nb in fully annealed condition
External diameter of cladding	13.6mm
Wall thickness of cladding (min)	0.825mm
Diametral gap between fuel and cladding	0.18 - 0.38mm
Fuel enrichment	2.0%
Height of fuel pellet	15.0mm
Diameter of fuel pellet	11.5mm
Volume of indentation on pellet	3%

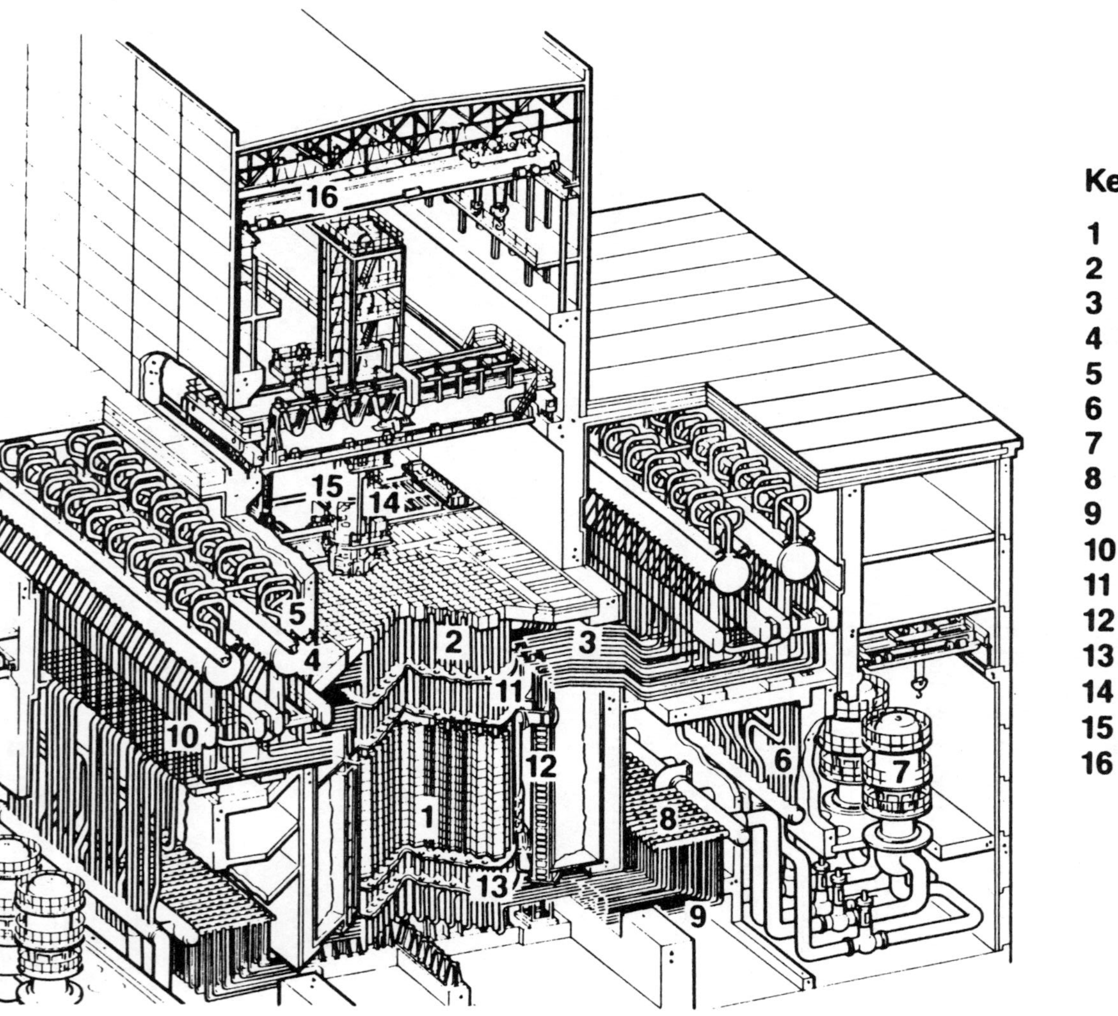

FIG. 1 SECTIONAL VIEW OF RBMK - 1000 REACTOR

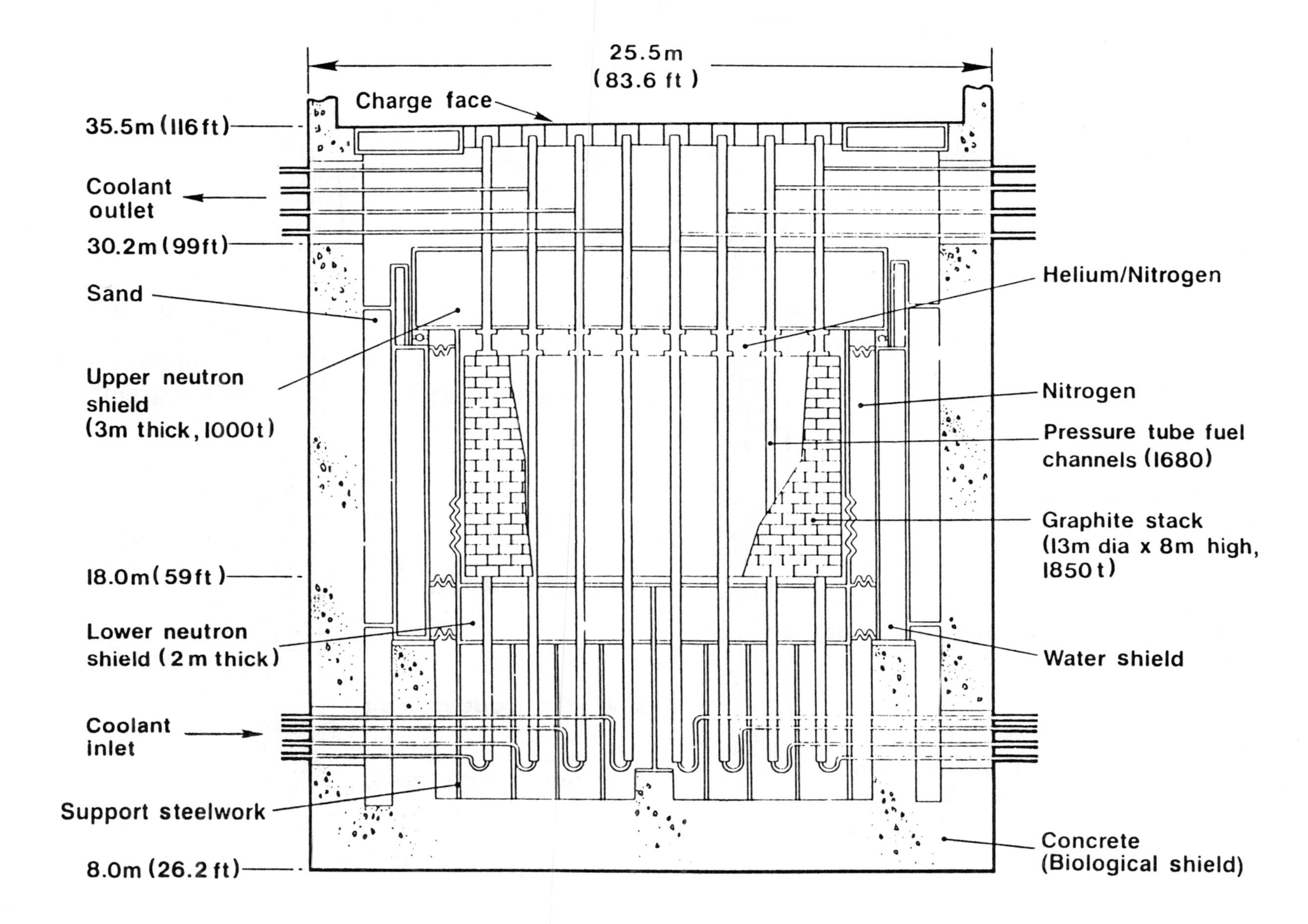

FIG. 2 SECTION THROUGH RBMK REACTOR

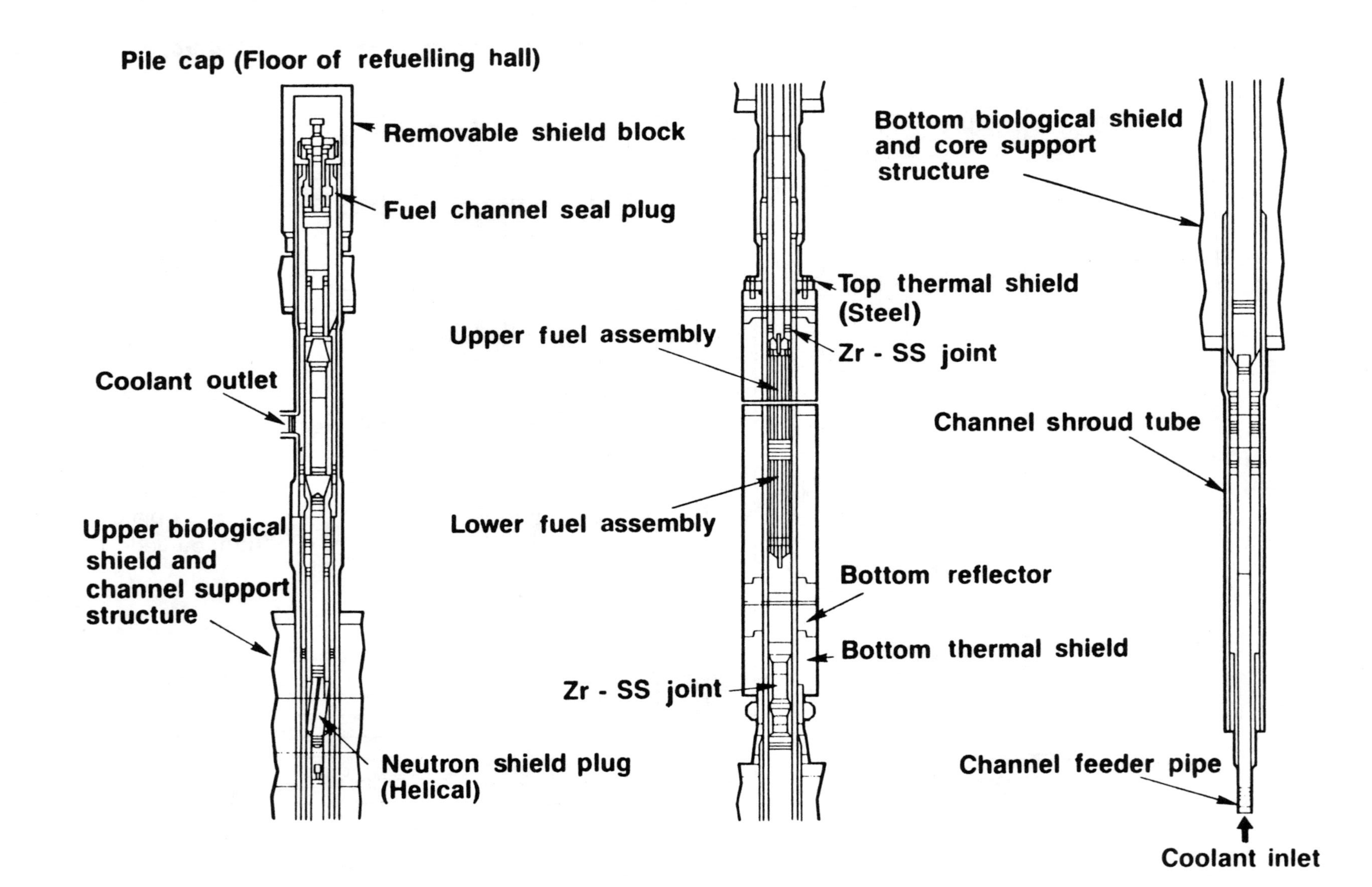

FIG. 3a FUEL CHANNEL

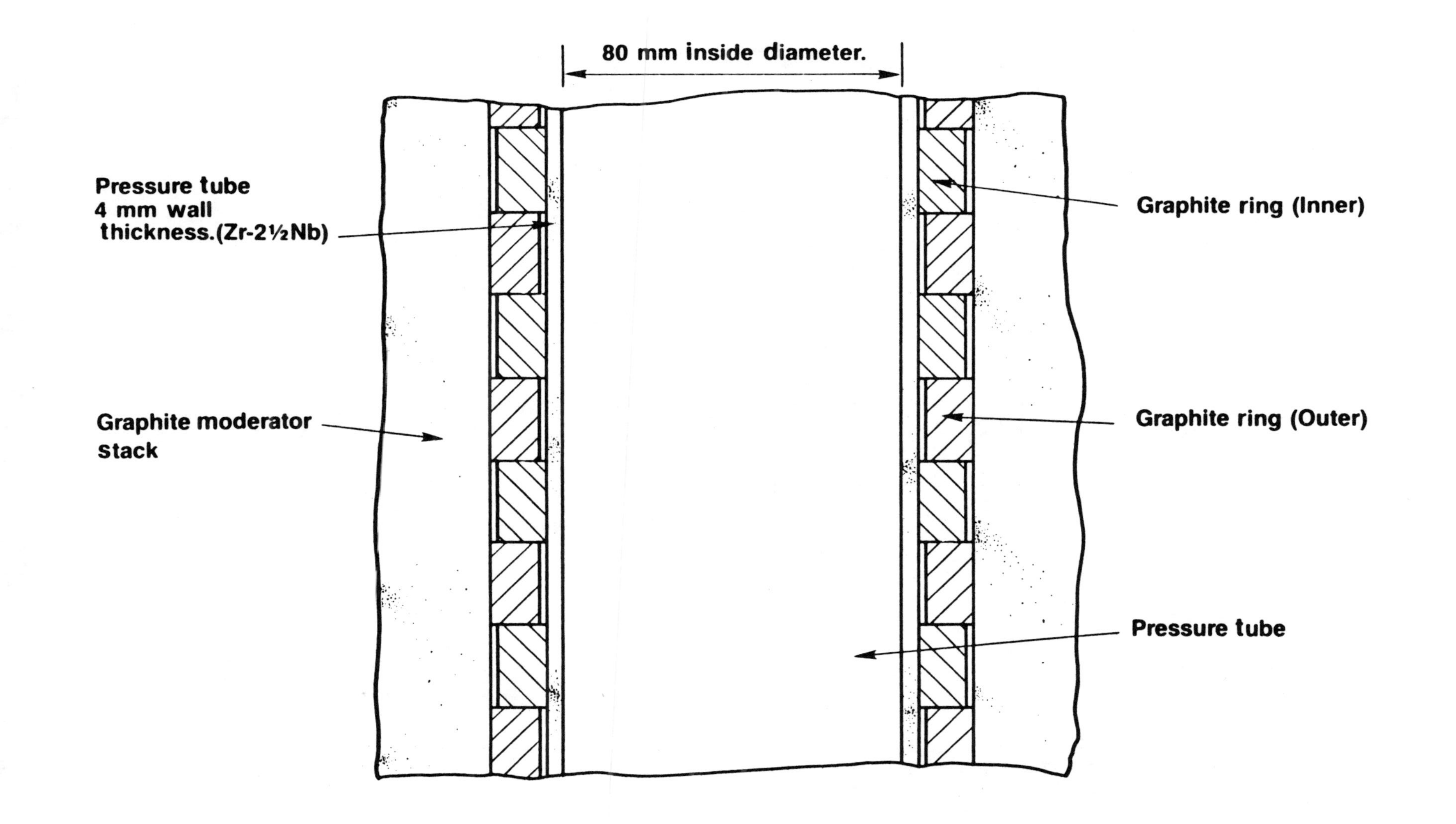

FIG. 3b ARRANGEMENT OF PRESSURE TUBES IN REACTOR CORE

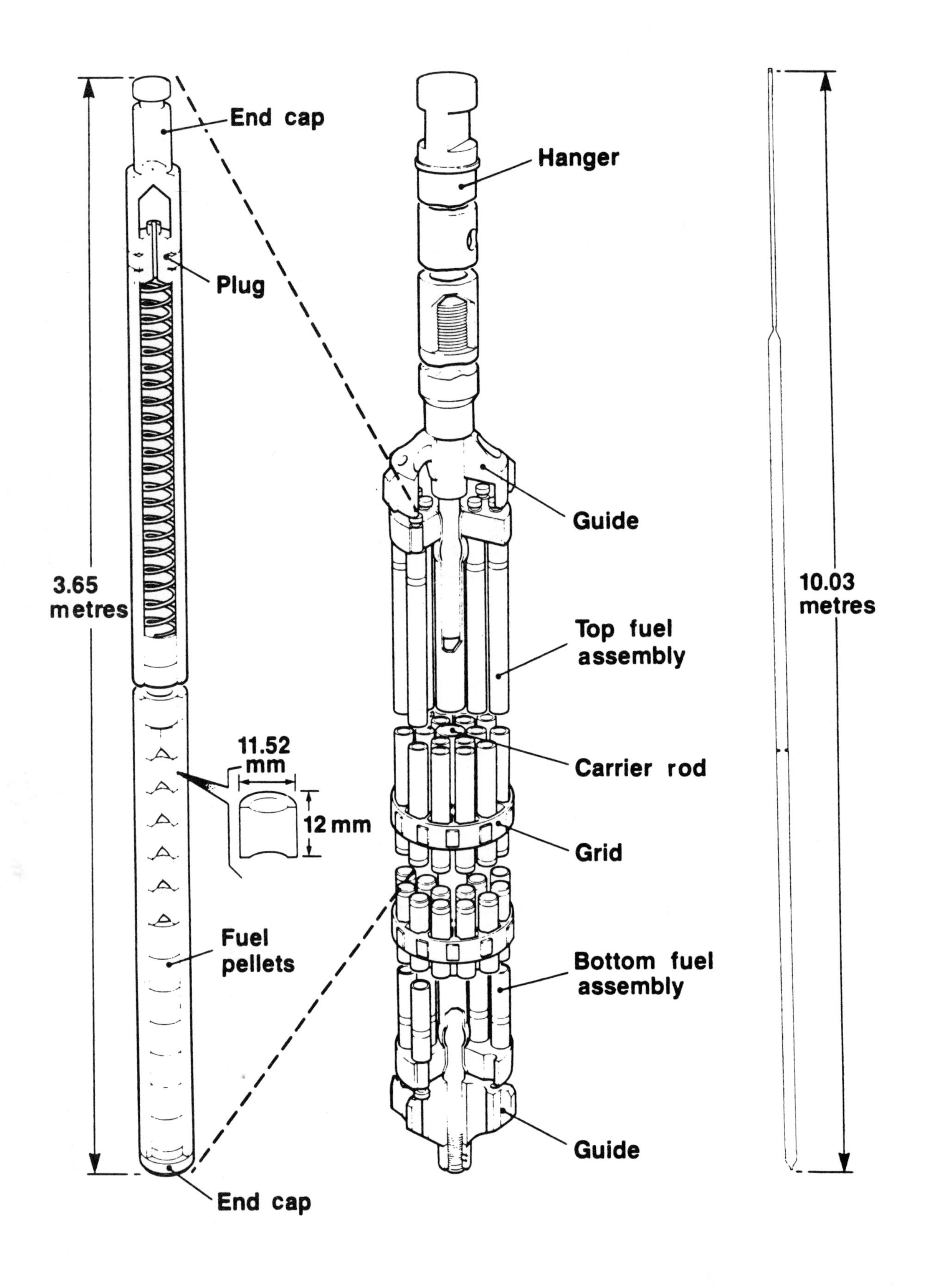

FIG. 3C RBMK FUEL ASSEMBLY

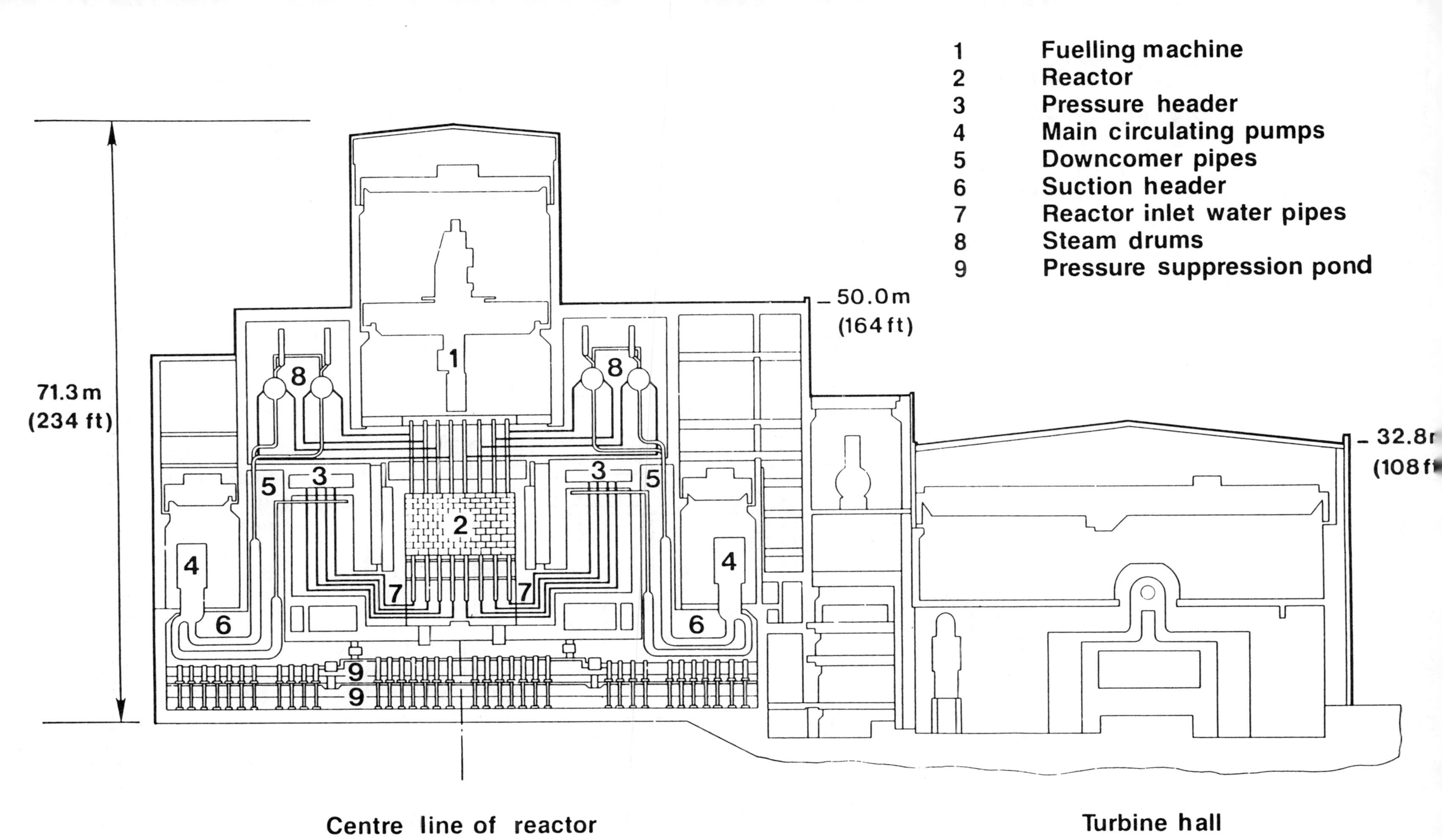

FIG. 4 CHERNOBYL - 4

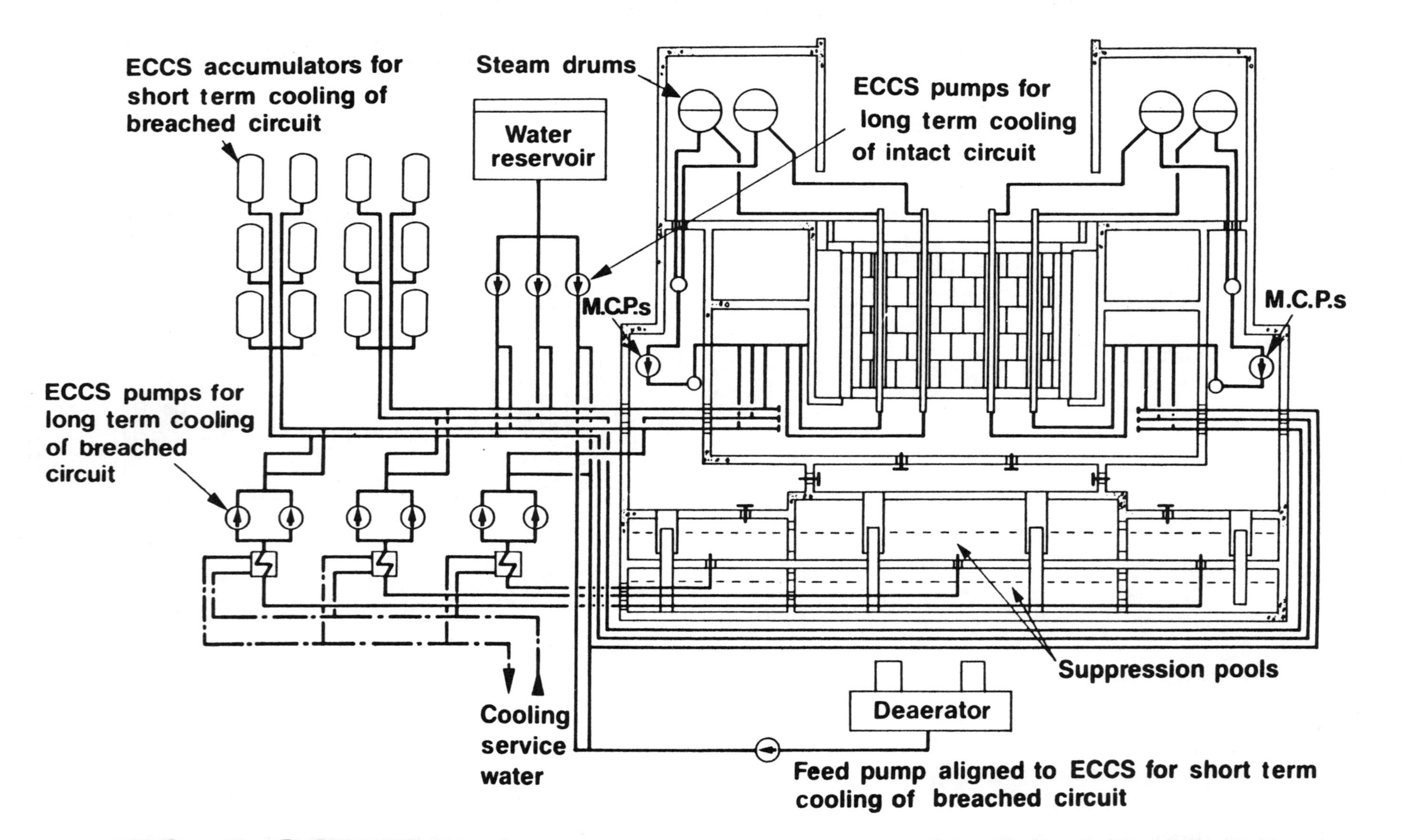

FIG. 5 SCHEMATIC DIAGRAM REACTOR COOLING CIRCUIT AND EMERGENCY CORE COOLING SYSTEMS

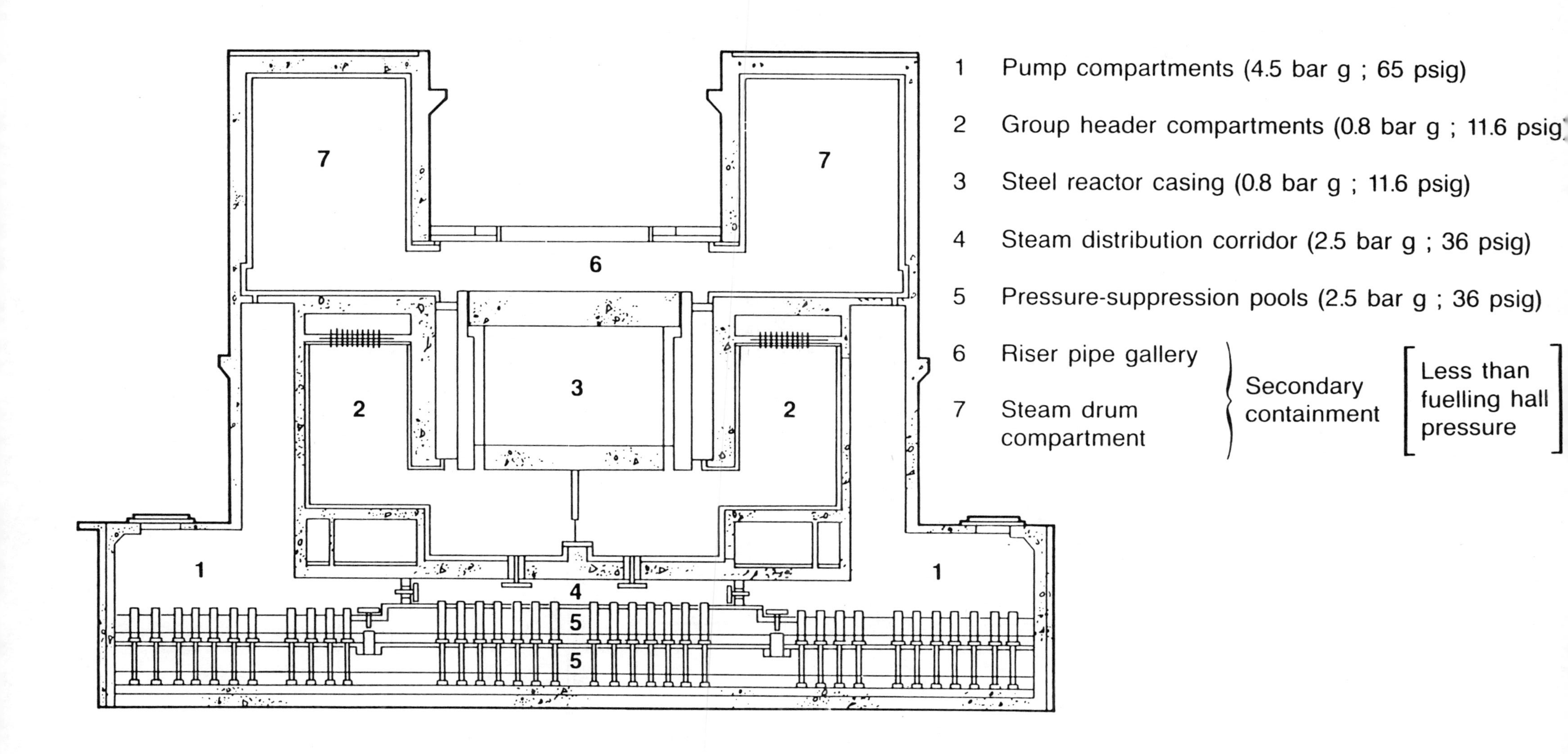

FIG. 6 SCHEMATIC DIAGRAM OF THE ACCIDENT - CONFINEMENT SYSTEM (ELEVATION)

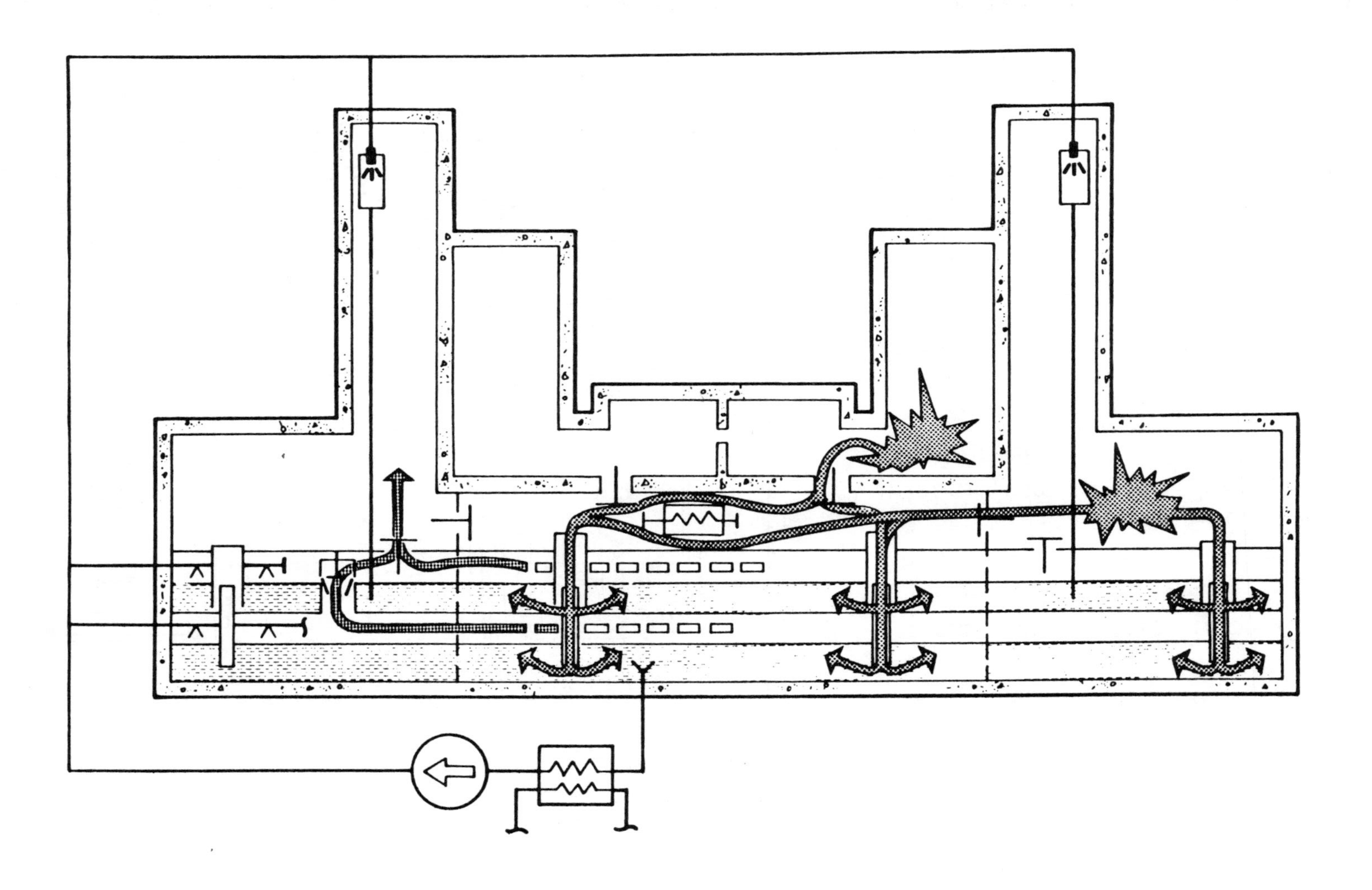

FIG. 7 VENTING ROUTES IN THE CONTAINMENT SYSTEM

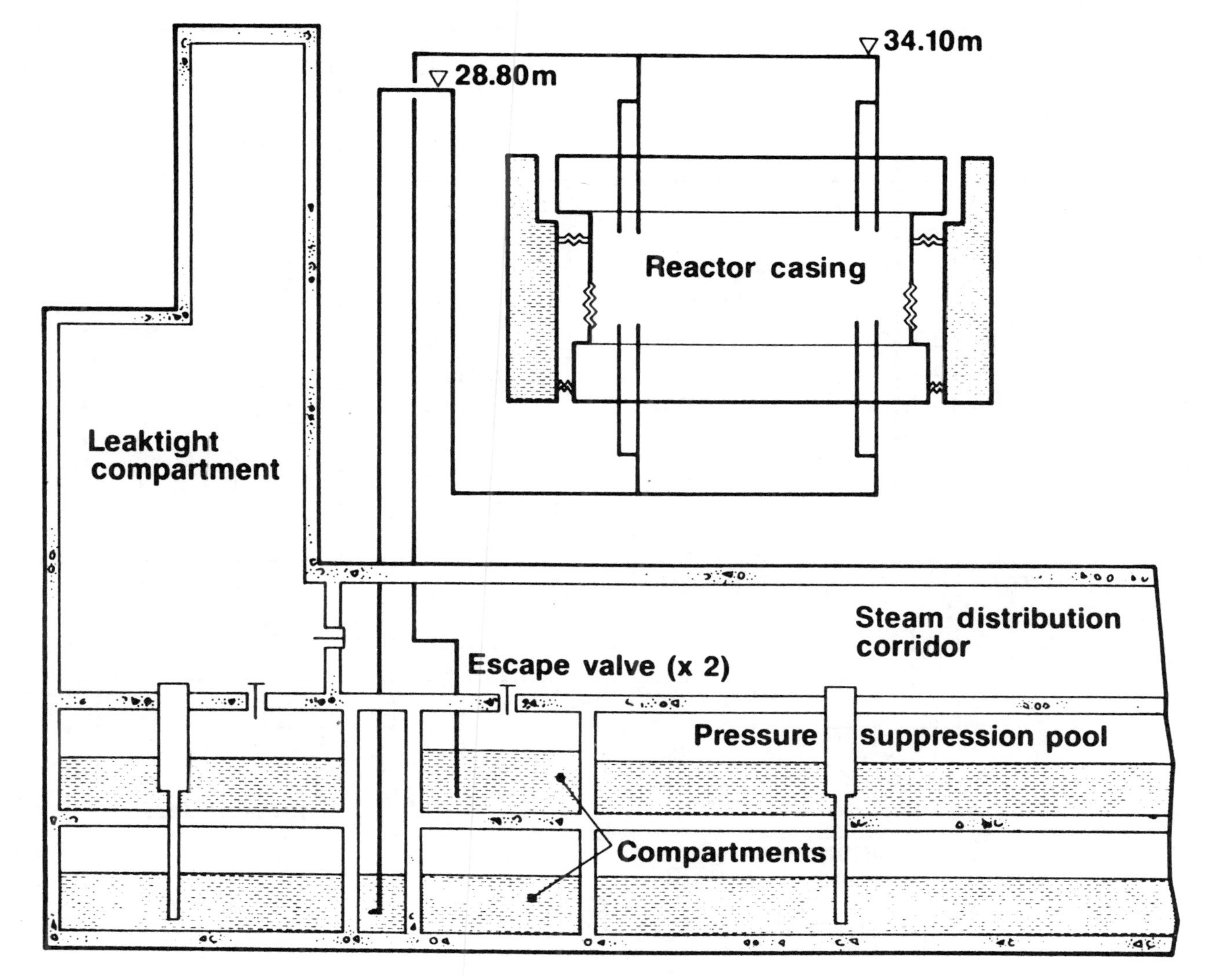

FIG. 8 SYSTEM TO PROTECT THE REACTOR CASING FROM EXCESS PRESSURE

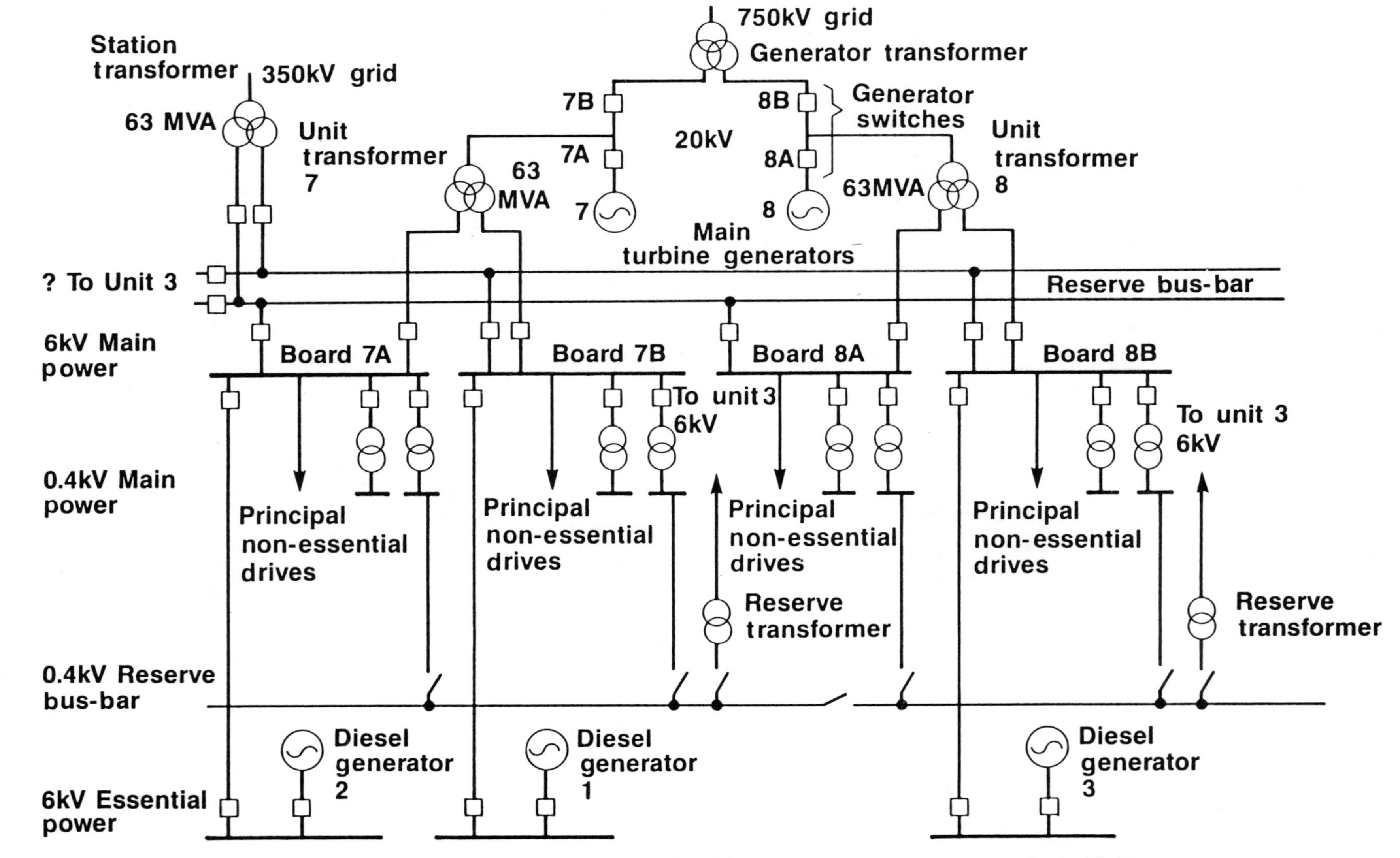

FIG. 9 MAIN ELECTRICAL SYSTEM

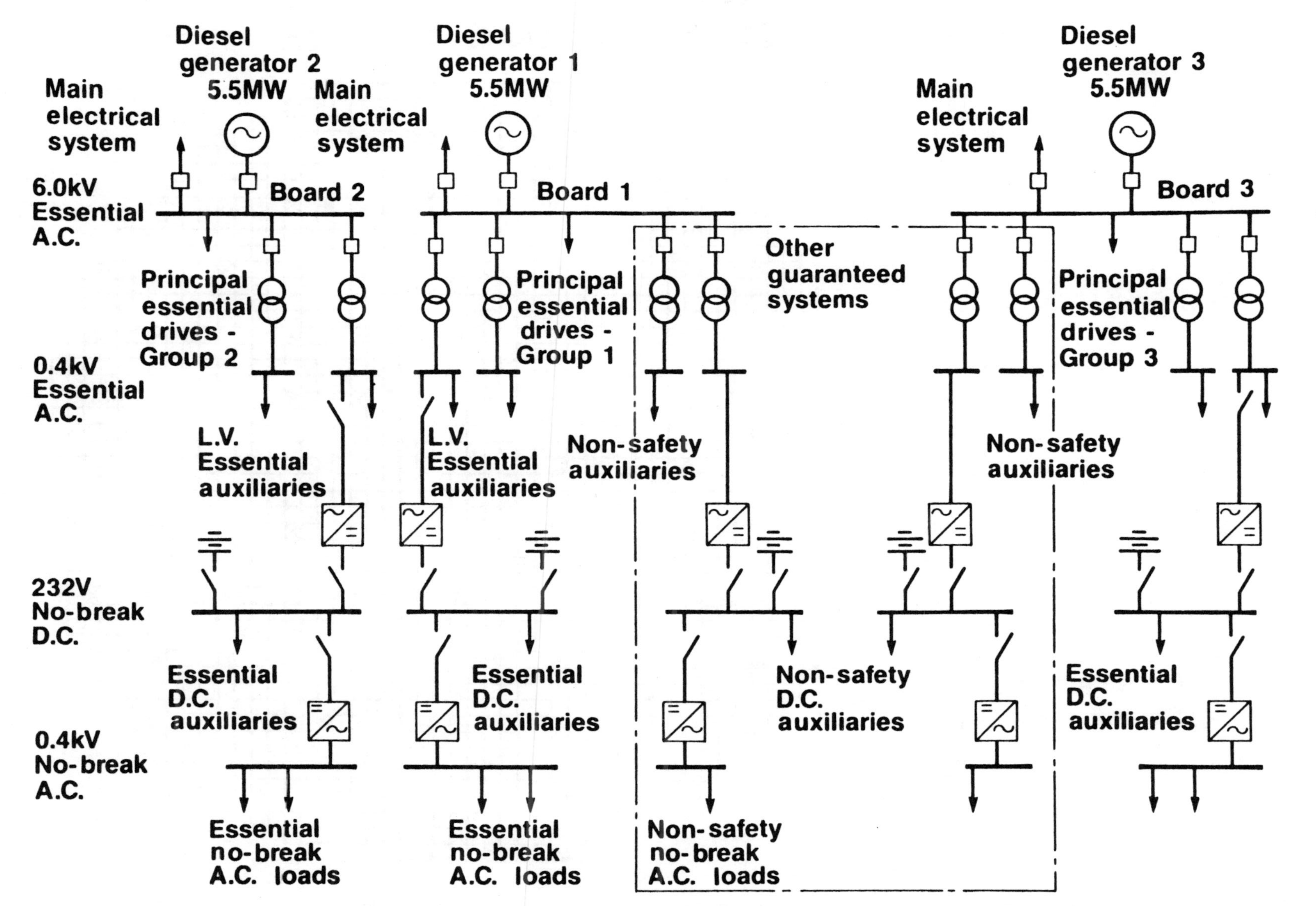

FIG. 10 ESSENTIAL ELECTRICAL SYSTEM

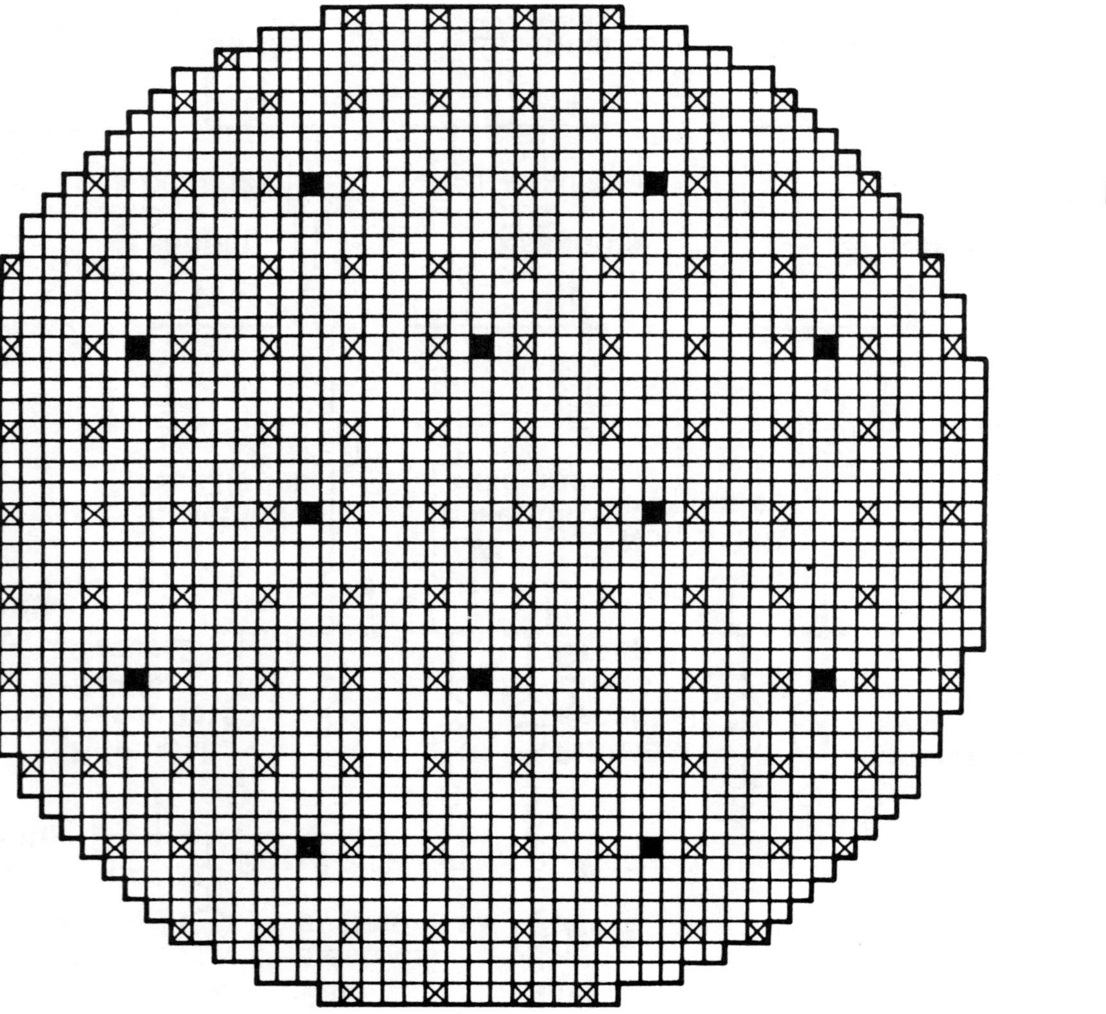

FIG. 11 POWER DISTRIBUTION MONITORING DETECTORS

APPENDIX 2.1 - PHYSICS CHARACTERISTICS OF RBMK

A2.1 Introduction

The object of this appendix is to review in a simple manner the performance characteristics of the RBMK reactor which are relevant to an understanding of the Chernobyl accident. It will be shown that, as in other reactors cooled by boiling water, the hydrogen atoms in the coolant have an important influence on the course of the neutron chain reaction. If there is some disturbance to the steam production in the core there will be a feedback effect on the power output which can in principle be positive (destabilising) or negative (stabilising) depending on the details of the core design. This appendix therefore presents a qualitative discussion of the physics of the neutron cycle and of the heat removal processes in the boiling coolant, so that the factors influencing the sign and magnitude of the feedback effect can be understood. It will be shown that the reactor was in a particularly unstable state immediately prior to the accident.

A2.2 The Operating Principle of a Thermal Reactor

The element uranium is found in nature as a mixture of isotopes with the approximate proportions 99.3% U^{238}, 0.7% U^{235}. Nuclear fission can only be induced in U^{238} by relatively fast moving neutrons with energies of about 1 MeV and upwards. The dominant effect of U^{238} is the absorption of neutrons. On the other hand fission can be induced in U^{235} by neutrons of all energies and indeed the process is particularly easy to induce with slow moving neutrons.

A thermal reactor is designed to exploit the above characteristics by arranging for the fast neutrons emitted in fission to be slowed down to an energy where there is a sufficiently high probability of inducing sufficient new fissions in U^{235} to produce a self sustaining chain reaction. The slowing down is produced by allowing the neutrons to collide elastically (ie to make billiard ball type collisions) with the atoms of a light element in a medium known as the "moderator". This process is continued until the neutrons come into a state where their motions are in approximate thermal equilibrium with the atomic motions of the atoms of the moderator. Such neutrons are described as "thermal neutrons" and their probability of capture by U^{235} must be sufficiently high relative to the chances of capture by U^{238}, structural materials, or the moderator, for sufficient fissions to occur to continue the process.

The first aim of the nuclear designers is to achieve appropriate volume ratios and geometrical arrangements of materials so that a chain reaction can be sustained and the heat satisfactorily removed. It is evident that this task can be eased by artificially increasing the proportions of U^{235} in the fuel above the natural value. This process is known as enrichment.

A2.3 Essentials of the Neutron Cycle in RBMK

At Chernobyl the No 4 reactor contained 1661 fuelled cells with the RBMK standard geometry shown in cross-section in Figure A1. At the centre of the cell is an 18 rod fuel cluster. Each rod consists of UO_2 pellets enriched to 2%, clad in a zirconium-niobium alloy. Boiling water coolant flows upwards in the spaces between the pins. The spacing is preserved by grids mounted at intervals throughout the height of the channel. The fuel cluster is surrounded by a zirconium-niobium pressure tube which is in thermal contact with the moderator graphite filling the remainder of the cell. In addition to the fuelled cells there are other cells which contain control rods or fixed absorbers within the same overall size of moderator block.

In order to understand the feed back mechanism operating when operation is perturbed, it is necessary to consider the neutron cycle in an RBMK cell in a little more detail. The essentials may be summarised as follows:

a Fast neutrons with a mean energy of about 2.5 MeV are emitted from U^{235} fissions occurring within the fuel pins. These neutrons travel outwards through the cell. A few will cause fissions in U^{238} atoms but most will wander through the cell making elastic (ie billiard ball) scattering collisions with the atoms of the various materials present.

b Moderation (ie substantial slowing down) will be produced by successive neutron collisions with the light elements carbon (in the graphite moderator block) and hydrogen (in the boiling water coolant). Because of its far greater volume the graphite produces most of the moderation, but since hydrogen is a particularly effective scatterer of neutrons, the water coolant has a more significant effect than its small volume would suggest.

c In the neutron energy region between 6 eV to about 100 KeV there are strong neutron capture resonances in U^{238}. If a neutron undergoing slowing down wanders into an energy band corresponding to such a high probability resonance it is possible that it will undergo "resonance capture" and so be lost to the chain reaction. It is essential in a thermal reactor to have sufficient moderating material present to ensure that a neutron which scattered into the resonance energy band will have a high probability of being scattered out to a lower energy before it is captured. The so-called "resonance escape probability" must be high enough to sustain the chain reaction.

d Once the neutrons are thermalised they diffuse within the cell until they are captured by the fuel, moderator or structural materials. It is obviously essential that sufficient neutrons should be captured by the fuel to sustain the reaction.

The neutron cycle in an infinite lattice may be expressed by the four factor formula, ie

$$k_\infty = \varepsilon \, p \, \eta \, f$$

where k_∞ is the reproduction factor of an infinite lattice, ie the ratio of the neutron population in successive generations.

ε is the fast fission factor corresponding to the enhancement in the fission rate produced by fissions induced by fast neutrons in U^{238}.

p is the resonance escape probability

η is the number of neutrons emitted in new fissions per neutron captured in fuel.

f is the thermal utilisation, ie the fraction of thermal neutrons captured by the fuel.

In a finite reactor there is an effective reproduction factor k_{eff} (which must be unity in a steady state) defined as

$$k_{eff} = \varepsilon \, p \, \eta \, f \, L$$

where L is the non-leakage probability, ie the probability of a neutron not leaking out of the edge of the reactor.

A2.4 Optimisation of the RBMK Lattice Design

From the preceding discussion it can be seen that the nuclear characteristics of the reactor core will be strongly influenced by the relative volumes of the materials present. In particular an increase of the volume of graphite will increase the value of the resonance escape probability p. However, graphite is also a neutron absorber and this will tend to diminish f. There is therefore, an optimum degree of moderation which will give the cell its highest reproduction factor. In practice this will also correspond approximately to the lattice which will give the longest fuel burn-up for a given enrichment of fuel. The designers of the RBMK treated the economy in the use of U^{235} enrichment as an overriding objective and chose a graphite volume very close to the optimum value.

A2.5 Void Coefficients of Reactivity

The amount of steam produced in the channels of an RBMK reactor obviously influences the neutron behaviour. The density of steam being much less than the density of water, it is usual to think of the steam production as being equivalent to producing voids in the cooling water, ie lowering the effective density. The discussion is usually conducted in the Western World in terms of a void coefficient of reactivity defined by the equation

$$k_v = \frac{\text{Percentage change in effective reproduction constant}}{\text{Percentage change in steam voids}}$$

NB Certain Soviet writers use a definition based on the absolute reactivity change which makes the quoted coefficient smaller by a factor of 100.

An increase of steam voids will produce two important effects in an infinite lattice, ie:

1 The reduction of hydrogen atoms in the cell will reduce moderation and therefore reduce the resonance escape probability. This effect always makes a negative contribution to the void coefficient.

2 Since hydrogen atoms are also neutron absorbers a reduction in their numbers provides less competition with the fuel for the absorption of thermal neutrons. The thermal utilisation factor f is therefore increased. This effect always makes a positive contribution to the void coefficient.

There is a third contributor to the void coefficient which, whilst unimportant in a large uniformly loaded reactor, will need to be referred to in the discussion below. In a real reactor of finite size, neutrons leak from the boundaries. The diffusion of neutrons towards the boundaries implies a net current flow. Such a current flow is impeded by the presence of neutron scattering material which includes the hydrogen atoms in the coolant. Thus when the effective coolant density is diminished by extra steam production, neutrons can flow more easily towards the boundaries and leak from the system. This provides a negative contribution to the void coefficient. Absorbers within the core can cause a similar effect by providing additional leakage boundaries.

Before leaving this discussion of void coefficients it is important to point out that in a real reactor the void coefficient is not a unique quantity. It will depend on the initial state of the reactor and the nature of the disturbance considered. This is because the effect of a steam bubble will depend on position in the reactor. If the bubble is in a region of high neutron flux, and therefore high local fission rate, it will obviously produce a proportionately larger disturbance in the total fission rate in the reactor. On the other hand bubbles near boundaries where the local neutrons are more likely to leak out are much less effective. The simple theory of a uniform reactor without reflectors shows the effect of a bubble to depend on the square of the local neutron flux, but in real non-uniform reactors the effect is more complicated.

A2.6 Stability of the RBMK

In the RBMK lattice the two large effects contributing to the steam void coefficient discussed above are of approximately equal magnitude and opposite signs. The magnitude and sign of the void coefficient then depends rather critically on the precise details of the core loading and on some smaller effects neglected in this discussion.

The Chernobyl Reactor No 4 was initially loaded with 2% enriched fuel. If all of the 1661 channels had been loaded at the start the reactor would have had a large excess of reactivity which could not have been held down by the control absorbers. Some of the channels were therefore loaded with fixed neutron absorbers instead of fuel. The extra neutron absorbers increased the total neutron absorption in the core and therefore reduced the relative importance of absorption by hydrogen atoms expelled when steam was formed. This diminished the positive contribution to the void coefficient associated with the thermal utilisation factor f. Moreover, the absorbers being localised in particular channels, also acted like internal boundaries by inducing neutron current flows towards them. This made a negative contribution to the void coefficient. The combined effect was that the overall void coefficient for the start-of-life of the Chernobyl No 4 reactor was negative.

As operation proceeded the fuel began to burn up and fixed absorbers were progressively removed to compensate for the lost reactivity. Eventually replacement of fuel elements with the on-load charge discharge machine was also started. At the time of the accident, this process was well on the way to reaching the eventual equilibrium condition and the void coefficient had become positive. This was a direct result of the designers' decision to pursue fuel economy by providing a close to optimum volume of graphite moderator. In this case, a reduction in the number of hydrogen atoms in the core arising from additional steam production had only a small effect on total moderation but represented a more significant expulsion of absorber.

The designer relied on the so-called Doppler coefficient to cancel the positive feedback provided by the void coefficient and so stabilise the reactor. The Doppler coefficient is associated with the resonance absorption by U^{238} atoms in the fuel. As power increases the fuel rises in temperature and the thermal motions of the U^{238} atoms are enhanced. This increases the effective width of the band of neutron energies that can be captured by a particular resonance. The resonance escape probability falls and reactivity diminishes. The coefficient is always negative and opposes the effect of the positive void coefficient. At normal full power the Chernobyl No 4 reactor was marginally stable with the negative feedback from the Doppler coefficient overriding the positive feedback from the void coefficient. As power is reduced, both coefficients increase in magnitude but the void coefficient can then predominate giving a net positive feedback effect at low powers. For this reason the designers imposed an operating rule that forbade sustained operation below 20% power.

A2.7 Heat Removal from the Moderator

The designers of the RBMK decided to seek the highest possible thermal efficiency in their reactor by recovering the heat from the graphite moderator. This heat arises partly from the kinetic energy dissipated when neutrons are slowed by collisions with carbon atoms, and partly from gamma radiation emitted during fission itself or from the radioactive decay of fission fragments. This moderator heat amounts to approximately 5% of the total heat output. In the RBMK the heat is transmitted by conduction to the pressure tube through special graphite rings (see Figure A1), and thence into the main boiling coolant. This implies that the graphite moderator blocks must run hotter than the coolant. It also means that, since the peak graphite temperature should not be too high ($\sim$ 700°C) the thermal diffusion paths within the graphite must be relatively short. Very roughly therefore, neutron physics considerations having fixed the volume ratios of the various core materials, the need to remove the heat from the graphite fixed the scale. The upshot was that a very large number of relatively small diameter channels were needed to produce the output even when they were made about twice as tall as is usual in other boiling reactors.

The large physical height of the channels in relation to the neutron migration length, ie the root mean square distance travelled between birth of a neutron in a fission and its eventual capture, mean that the coupling between top and bottom of the core is weak, the axial power distribution becomes floppy and needs to be controlled carefully.

A2.8 Heat Removal from the Fuel

In normal operation cooling water enters the pressure tubes in a subcooled condition, ie at a temperature below the boiling point at the local pressure. At some distance up the channel (20–30% of the height) boiling commences. In the upper parts of the channel the flow settles into the so-called annular flow regime. Thin water films travel upwards over the surfaces of the fuel pins. Steam flows more rapidly in the spaces between the pins carrying water droplets with it. So long as the films are maintained the surfaces of the fuel elements are in good thermal contact with water at the boiling point (ie about 300°C) and will be well cooled. At any level in the core the film will have a thickness representing the balance between

a Evaporation loss produced by the heat flow from the fuel.

b Deposition gain as water droplets carried by the steam flow collide with the film and are absorbed.

c Entrainment loss as water drops are ripped from the crests of waves established on the film surface by the steam flow.

In general as the coolant travels up the channel the film thins as water is progressively turned to steam. Eventually near the top of the channel as the rate of heat production falls away with the neutron flux, the film thickens again due to capture of the remaining water droplets. Somewhere towards the top of the channel the film is at its thinnest.

If the normal operational condition is disturbed by either increasing the total power input, or reducing either coolant flow or subcooling, the water films on the fuel pins may break. This is the phenomenon called "heat transfer crisis" by the Soviets or "dryout" in the West. When it happens the heat transfer at the fuel cladding surface becomes far less efficient and the clad temperature rises from 300°C to 1000°C or beyond. This greatly increases the rate of corrosion of the zirconium alloy and can only be tolerated for a very short time or severe fuel damage will occur and render the fuel unfit for further service.

RBMK reactors are designed to allow the operators control over the margin to dryout in every channel so that the power output can be optimised. The flow in individual channels may be adjusted and the "floppy" axial power shape trimmed by moving "short" control rods to heights where power peaks need to be suppressed. Special instruments provide an estimate of the margin to dryout in each channel.

A2.9 Control of the RBMK Reactor

In order to understand the physics of events occurring immediately before the accident it is necessary to consider the action of the control system. This is used to maintain the reproduction factor of the reactor at unity if steady operation is required or close to unity if controlled increases or decreases of power are required. The process is made easier by the fact that approximately 0.7% of fissions occurring in U^{235} result in fission fragments which undergo radioactive decay to excited nuclei which "boil off" so-called delayed neutrons. The half lives of the radioactive decays range up to about 1 minute and these slow down the rate at which successive generations of neutrons in the chain reaction can multiply so long as the reproduction factor is less than 1.007. However, if the reproduction factor is greater than 1.007 the neutron population will multiply at a rate governed by the neutron lifetime which is approximately 1 millisecond. This is referred to as "prompt criticality".

Irradiated fuel also contains Pu^{239}, which yields a smaller delayed neutron fraction than U^{235}. This reduces the delayed neutron fraction of irradiated fuel to about 0.5% in RBMK reactors and therefore reduces the reproduction factor at which prompt criticality would occur to 1.005.

It is evident that control absorbers in reactors should be capable of quick insertion to correct a fault. In the Chernobyl RBMK, the control rods travel in water cooling tubes set in the graphite. They do not travel very freely and require approximately 20 seconds for full insertion from the fully withdrawn position. The high core height combined with the "floppy" loosely coupled axial power distribution meant that the absorbers had to travel an unusually large distance to produce significant effects. The designers were well aware of this and produced an operating rule which required a specified number of absorbers to be poised part way into the core. In this position they would "bite" quickly and develop an adequate effect in less time.

Since the first analysis of the causes of the accident was published by the Soviets in Vienna, further analyses and more information on design details have highlighted an additional feature of the reactivity aspects of control rod insertion. This feature called "positive scram" could have contributed to the severity of the accident. This comes about in the following way.

In the RBMK reactor, if the boron carbide control rods simply displaced water, itself an absorber of neutrons, then the effectiveness of the rods would be reduced. This is avoided by the attachment of non-absorbing graphite rods or "followers" to their ends which displace the water which would otherwise fill the channel when the rods are withdrawn. This maximises the increase in neutron absorption when the rods are reinserted and therefore makes them more effective at shutting down the chain reaction.

Due to height restrictions within the reactor these graphite rods are shorter than the fuelled region of the core, by 2.5 metres. A dangerous situation could therefore arise in certain circumstances as the displacement of water (a good absorber of neutrons) by graphite (a non-absorber) could result in a speed up in the chain reaction. The effect is analogous to the existence of a positive void coefficient, except that in this case water is physically displaced from the core by the rod follower, rather than boiling and turning into steam. It should be noted, however, that the displacement of water from a control rod position results in an enhanced change in absorption (f) relative to resonance escape (p) compared to voidage in the fuel channels. The potential for positive scram would still exist even if the void coefficient were negative, given the control rod/follower geometry at the time of the accident.

Under normal circumstances, few if any of the control rods would have been fully withdrawn and even if they had been, the entry of the absorber rods at the top of the core would have dominated the overall change of reactivity and shut the reactor down. In the accident, however, almost all the rods had been withdrawn from the core and xenon poisoning in the centre of the core could have allowed the top and bottom of the core to become to some extent neutronically decoupled. The entry of the absorber rods at the top of the core might then have had reduced influence on events near the bottom of the core and the displacement of water by the graphite followers might have worsened the developing power surge. The magnitude of the excess reactivity introduced by this "positive scram" effect would have been very sensitive to the exact conditions in the core at the time of the accident and without more detailed knowledge of these an upper limit of 1.5 β was assessed in the UK analyses. Nevertheless it does appear likely that the attempt to shutdown the reactor did result in a 'positive scram' and that this contributed to the power surge. This basic conclusion has now been confirmed by the Soviets [see the paper by Asmolov et al, IAEA International Conference on Nuclear Power Performance and Safety, which quantified the magnitude of this effect as 0 - 1.5 β based on reactor measurements].

A2.10 Comments on Events Preceding the Accident

The Chernobyl No 4 reactor was completing a period of full power operation and being progressively shut down for maintenance in the period preceding the accident. Power reduction would have been initiated by a partial insertion of control absorbers. The operators were however, requested to pause at an intermediate level because continuing output was needed to meet the load. The build-up of Xenon poison then became a factor in the situation. One of the common fission fragments is I^{135} which decays to Xe^{135} with a half life of 6.6 hours. Xe^{135} has a very large neutron capture cross-section. In steady operation, the creation and destruction of Xe^{135} come into equilibrium and the Xe^{135} population stabilises at a constant level. When power is reduced to a new lower steady level, the destruction rate of Xe^{135} is reduced and a transient increase of the Xe^{135} population occurs, before it stabilises at a new lower level. Thus for a period following a reduction of power, there is more total Xe^{135} "poison" in the reactor which must be compensated by withdrawing control absorbers. There was thus a trend at work moving the operators towards the operating limit governing minimum numbers of rods inserted in the core to give adequate "bite".

The situation was exacerbated by the operators leaving the controls in a "power reduction" mode which took the power well below the 20% operating limit. There was now very little steam void in the core. In view of the positive void coefficient the loss of the positive steam void reactivity led to the need for further control rod withdrawals and the situation was now well outside the rules.

Just before the experiment began the operators applied a greatly increased feedwater flow and managed to stabilise the power level at about 6%. This cooled the channels and reduced voidage still further. At this stage the normal "annular flow" flow regime described above might not have been properly developed and the channels may have been operating in a "bubbly flow" regime. This is particularly likely with the very high flow produced by turning on the extra pump in each circuit. In the bubbly flow regime steam bubbles form on the heated channel walls and are swept away by the water flow. In any case the change back to the annular flow regime when the excessive feed flow was reduced could have been a factor in triggering the accident.

A2.11 Comments on the Accident Itself

The sequence has been described fully elsewhere, the object here is to comment on the physics. As the coolant flow to the core inlet decreased, the rate of boil-off rose, increasing the volume of steam in the core. This tended to increase the reactivity of the core due to the reactor's positive void coefficient, but the automatic control rods initially compensated for this effect. When they could no longer do so, the power of the reactor started to rise. At the low power at which the experiment was being conducted, an increase of power caused a further rise of reactivity as the void effect on reactivity exceeded the counteracting fuel Doppler effect. In other words, the power coefficient of the reactor was positive. The increased core voidage spread from the end of the channel to regions where the reactivity effect was greater and also increased the flow resistance still further reducing flow. Thus, while the neutron physics characteristics drove the power up, the hydraulics characteristics accelerated the flow run down. As explained in A2.10, the phenomenon of positive scram to which this design was susceptible could also have added to the reactivity increase caused by the positive feedback effect. Coupled with the long time required to drive the control rods in, and the extent to which they had been withdrawn, this lends credence to the idea that the act of activating the shutdown rods could have been the final trigger for the massive reactivity insertion. Although the power coefficient would have become negative by the time the normal full power level had been reached, coolant voiding and positive scram had increased the reactivity so much that the fuel Doppler effect could only terminate the power surge if the fuel temperature rose to values very much greater than those of normal operation. "Dryout" must therefore have occurred on the fuel surfaces, leading to a virtually total loss of local cooling. The reactor became prompt critical and the very high heat input then led to disruption of the fuel pellets and explosive generation of steam by the small fragments giving up their heat to the cooling water.

A2.12 Comments on Remedial Measures

In the short-term the Soviets proposed to fit all control rods with limit switches ensuring a minimum insertion of 1.2m. However, this was found to lead to a distortion of the axial neutron flux profile that required the power to be reduced by 10 to 15%. It was therefore abandoned in favour of a modification to the linkage between the control rods and followers that would allow the control rods to be fully withdrawn without raising the followers above the bottom of the core. This prevents any possibility of a "positive scram" effect.

They also propose to keep the equivalent of 70-80 rods within the core. This should prevent the development of a control rod configuration that violates the demand for a minimum "reactivity reserve" (ie potential increase of reactivity if all control rods were withdrawn). The preceding discussion shows that this should also reduce the void coefficient and the overall protection against design basis faults will undoubtedly increase. The total absorption of the core is increased and therefore the percentage change in absorbtion associated with the replacement of a given amount of water by steam is less, ie a positive contributor to the void coefficient is reduced. Furthermore, the rods induce internal neutron current flows which increase as water changes to steam giving a negative contribution to the void coefficient.

It is obvious that the short-term measures increase the total neutron absorption within the core which must be compensated in some way. This can be done by discharging the fuel with a shorter burn-up which implies an increase in the frequency of refuelling channels on load. This has an adverse effect on the generating costs since less energy is extracted per tonne of fuel loaded. It is not surprising that the Soviets therefore propose, as a second stage remedial measure, to increase the fuel enrichment from 2.0% to 2.4%. This will give extra reactivity to compensate for the loss of burn-up. Since the extra U^{235} increases still further the total absorption of the core, there will also be a useful further reduction in the positive contribution to the void coefficient associated with the replacement of water by less neutron absorbing steam.

The measures discussed should greatly reduce the likelihood of a rapid transient increase in power in RBMK reactors. It is however essential, that to reach the normal standards applied in Western reactors, the Soviets persist with their intention to add a fast acting shutdown system capable of arresting any unforeseen transient which might develop.

A2.13 General Comments

This appendix has shown that in the pursuit of extracting the maximum possible energy from a given supply of U^{235} the Soviets arrived at a design with unique characteristics. In particular, once the plant had been running for long enough to approach the equilibrium rate of refuelling, the coolant void coefficient was positive. As a result, complicated operating rules had to be followed to ensure stability. However, no interlocks or other restraints were provided to prevent the operators from breaking the rules with the consequences described elsewhere in this Report.

It has been shown at least qualitatively that the remedial measures proposed by the Soviets are likely to be effective in preventing a similar power excursion accident in the future.

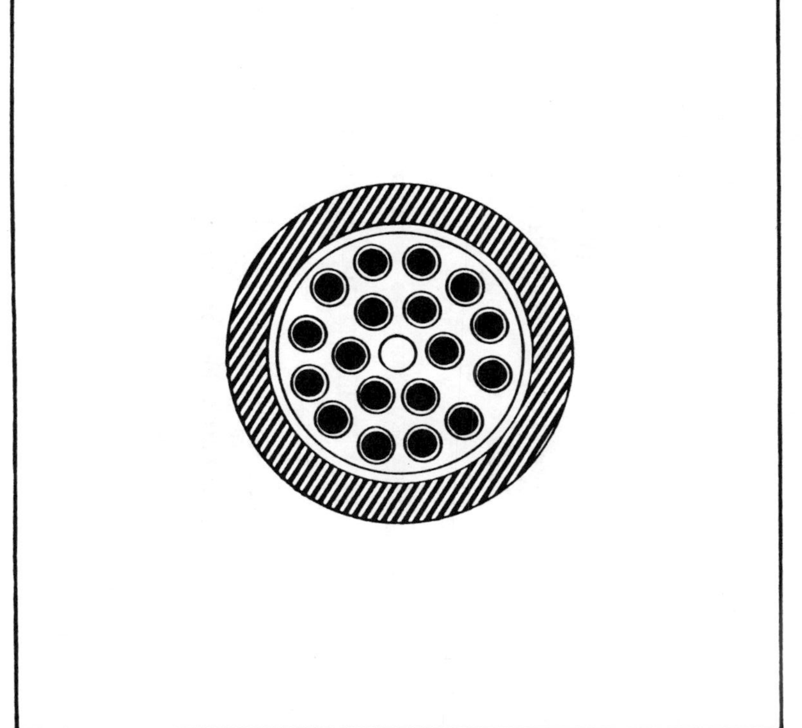

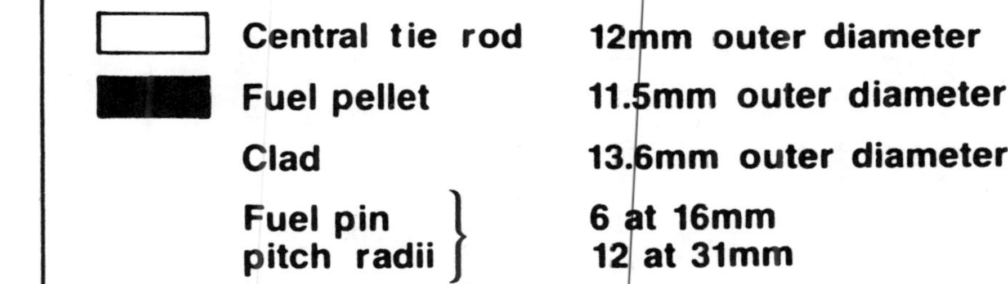

KEY

(white)	Central tie rod	12mm outer diameter
(black)	Fuel pellet	11.5mm outer diameter
	Clad	13.6mm outer diameter
	Fuel pin pitch radii	6 at 16mm 12 at 31mm
	Pressure tube	80mm inner diameter 88mm outer diameter
(hatched)	Graphite conduction rings	114mm outer diameter
	Lattice pitch	250mm square

FIG. A1 SECTION THROUGH FUELLED LATTICE CELL

SECTION 3: NNC SAFETY RESERVATIONS

Introduction

In 1975 active measures were in hand to promote relations between the USSR and the UK. Nuclear power was judged to be a suitable area for fruitful interchange. A British Nuclear Forum Delegation, including NNC staff, therefore visited the USSR in October/November 1975 and there was also a return Soviet visit to the UK. The main objective of the visit was to see what could be learned from the large effort that the Soviets were making in the design and construction of the RBMK pressure tube type of reactor. At this time the UK was engaged on the design of the Steam Generating Heavy Water Reactor (SGHWR), also a direct cycle pressure tube reactor, as a commercial power producer. There were some superficial similarities between the two concepts, and collaboration between the two projects was considered to be potentially beneficial.

A report (Ref 1) was produced as an internal review over ten years ago by the Nuclear Power Company (a predecessor of the NNC).

In the report reservations were expressed on a number of safety related aspects of the RBMK design. Although the report was not passed to the Soviets they were made aware of the reservations during meetings both in the USSR and in the UK and it is of interest that design improvements in the later RBMKs address some of these reservations. The safety topics from the NNC report are summarised in this section together with comments on the improvements incorporated in the Chernobyl design. Despite these improvements the Soviets acknowledged in Vienna that the Chernobyl design had fundamental shortcomings and these are noted where appropriate.

The Soviets now plan to make further modifications which are discussed in Section 4.

Safety Aspects

i The lack of a direct in-core spray emergency core cooling system: Ref 1 noted that the basis of the Leningrad RBMK design was that "gross failure of large bore pipes is incredible". It was also recognised by the

Russians that "stagnation could occur in some channels during a LOCA in the Leningrad reactor but this is considered so unlikely to be prolonged that it is just a risk to be taken".

It was the NNC view then that "it would almost certainly be necessary to adopt the SGHWR principles" ie that an ECCS system should be provided that sprayed water directly into the fuel channels, in parallel with the main circulation.

It was noted that a system of accumulators had been developed for later reactors and the Chernobyl design is provided with 2 x 50% accumulator systems to cater for the short term cooling following the fracture of the largest diameter pipework. A further 50% is obtained from the main feed pumps. The system would still not deal with stagnation.

The complex piping system was one of the design shortcomings highlighted by the Soviets at Vienna.

ii The lack of full containment appropriate for a water cooled reactor: The absence of a pressurised containment was noted in Ref 1. The building structure was similar to that of gas cooled reactors in the UK. It was also noted that "pressure retaining containment structures are provided on the SGWRs to prevent the release of radioactivity to the environment following potential LOCAS up to and including the maximum credible LOCA (usually a primary circuit pipe breach)."

It was concluded that it is unlikely that any type of water reactor would now be built in the UK without a substantial containment building.

It was also observed in Ref 1 that "pressure suppression arrangements are being studied and may be used at some stage in future designs." In the Chernobyl design certain parts of the primary circuit are surrounded by leak tight compartments which vent to a suppression pool under the reactor. However, the pipework above the reactor up to the steam separators, the separators and a short length of the downcomer pipework is not part of the containment. For the regions that are contained, the close proximity of the primary circuit pipework makes propagation of failure a potential cause of exceeding the design pressures of these compartments. Particularly vulnerable in this respect is the reactor cavity if more than one pressure tube should fail.

For the above reasons this aspect of the Chernobyl design is still considered to be unsatisfactory.

iii The mechanical instability of the graphite core:
With respect to seismic loading the following was noted in Ref 1.

"The graphite columns will normally be maintained vertical by the pressure tubes with only small lateral forces. However, in a UK context seismic loading would have to be considered.

In principle lateral seismic forces could be restrained either by:

a Each channel tube carrying the lateral acceleration forces caused by the brick column, or

b The channel tubes and graphite columns bending laterally until the accumulated interbrick gaps are closed and the lateral load can be supported by a side restraint structure.

For the ground acceleration likely to be specified for the UK designs, each of the above mechanisms would give rise to severe bending of the pressure tubes. They need not necessarily fail but it might be difficult to make a satisfactory safety case."

There has been no change in this in the Chernobyl design.

iv   Insufficient protection against failure of a pressure tube:
A related problem to (iii) is the potential damage arising from pressure tube failure. Ref 1 states:

"The venting between the pressure tube and the brick bore is extremely small and if a channel tube failed the pressure at the brick bore locally would approach system pressure even for only a limited pressure tube crack. A classical thick wall cylinder calculation indicates that the brick would just sustain a steady radial pressure load equal to system pressure but this would be difficult to justify for an irradiated brick with internal stresses. Furthermore the pressure build up at brick end face joints would lift the upper part of the brick column. Thus it is most unlikely that the water/steam pressure could be contained within a single brick column. When steam escapes from the single column the pressures at adjacent interbrick faces would be determined by the stack vent areas, the pressure tube hole size and the channel flow resistances. The situation would be worsened by the hot graphite evaporating the water in the escaping water/steam mixture.

It is believed that any credible brick gap venting arrangement would lead to excessive pressure in interbrick gaps for anything but a very small channel tube breach. These pressures acting over whole brick face areas would give rise to enormous radial and axial forces and make the restraint of the core almost impossible."

There has been no change in this in the Chernobyl design.

v   The reactor has a positive void coefficient:
The NNC calculations reported in 1976 predicted a void coefficient in the range 0.02 to 0.03 which is consistent with the value quoted by the Soviets for normal power operation. It was observed in Ref 1 that "this could result in spatial instability together with safety problems during faults unless the core is adequately subdivided as regards control and to minimise the area in which channel stagnation could occur." It was also noted that the implications for fault transients "would have to be checked for the full range of faults." At that time NNC was unaware that, on trip, the control

rods are motored into the core, taking some 15 to 20 seconds to reach full insertion. Had this been known NNC would have expressed more serious reservations on the shutdown system performance following faults.

This characteristic was highlighted by the Soviets at Vienna as one of the shortcomings of the design. In normal high power operation, the positive void coefficient is counteracted by the negative Doppler fuel temperature coefficient giving an overall negative power coefficient. In subsequent discussion with IAEA experts, the Soviets admitted the most important shortcoming of all, namely that because the void coefficient is so large and positive, the RBMK reactor has an overall power coefficient which is positive below 20% power under normal operating conditions. There is no engineered protection system to prevent operation in this unstable regime.

It was the existence of this positive power coefficient that led to the very rapid power excursion during the ill fated experiment.

vi Insufficient shutdown margin:
The observation in Ref 1 was based on calculations carried out by NNC assuming that the control rod absorber material would be boron steel. Subsequently it has been learned that the control rods contain a more effective absorber, boron carbide, therefore this comment is probably no longer valid.

vii The possibility of zonal instabilities and local criticality in the core:

In (v) above the positive void coefficient was discussed in relation to stability and local core voiding due to stagnation. A further destabilising influence, on a longer timescale determined by the graphite thermal capacity, is the positive moderator temperature coefficient. It was noted in Ref 1 that "gas cooled reactor analyses suggest that it would create significant stability problems and that it would have an adverse effect on fault transients." It would also "increase the difficulties in meeting the shut-down/hold-down reactivity requirements." While (vi) above indicates that the design should be satisfactory, calculations would be required particularly for proposed new operating conditions.

In discussion at Vienna it was learned that the Soviets vary the helium content of the reactor cavity gas to control the graphite temperature and hence the reactivity feedback via the positive temperature coefficient. This is a further complexity in the operation of the reactor.

One of the requirements of Ref 1 for in-core flux detectors to protect against asymmetric faults has been provided in the Chernobyl design.

Instability of the power density distribution in the core was identified by the Soviets as one of the design shortcomings.

viii No back-up to the control rods for reactivity shutdown:
By 1976 it had become an "established UK principle that two diverse means of shutting down the reactor should be provided." It was a criticism of the design then that no such system was provided. At the Vienna meeting it became evident that the omission was made even more critical by the fact that, on trip, the control rods are motored into the reactor and the rate of insertion of negative reactivity is dependent on the operator having kept sufficient reserve control rod capacity in compliance with operating rules. No protection system was provided to ensure compliance with this requirement.

ix The high temperature of the graphite core throughout the life of the reactor:
Due to the method of heat removal from the graphite by conduction to the channel wall the graphite maximum temperature is in the region of 700°C. Based on UK graphite data, NNC calculations in Ref 1 predicted that "towards the end of life the falling graphite conductivity increases maximum graphite temperatures well above 750°C", which would be "unacceptable".

The graphite reaction with air is exothermic. At these temperatures, it increases rapidly with temperature. Hence it would "be essential to exclude air from the graphite environment except during low temperature shutdown conditions."

The graphite reaction with steam is endothermic but to ensure an adequate graphite life, the helium/nitrogen mixture in the reactor cavity is passed through a clean up loop. The clean up flow appeared to NNC to be inadequate for the expected inleakage of steam based on UK pressure tube experience.

It was also noted in Ref 1 that "there is practically no scope for increasing the channel linear power rating." This is inconsistent with the fact that some later RBMK reactors have channel powers 50% greater than the Chernobyl plant.

The heat stored in the graphite was one of the disadvantages identified by the Soviets at Vienna.

Of the above reservations expressed in 1975, the main characteristic of the design that led to the incident was the positive void coefficient, which at low power resulted in an overall positive power coefficient. Pressure tube failures, the absence of full containment and high graphite temperature were design aspects that contributed to the scale of the accident.

Reference

1 NPC(R)1275 March 1976.
The Russian Graphite Moderated Channel Tube Reactor.

SECTION 4: POST-CHERNOBYL DESIGN CHANGES TO RBMK REACTORS

In Section 2, the design of the RBMK reactor has been summarised and a number of shortcomings have been highlighted by reference to UK safety standards. In particular, there are several aspects of the design which do not conform to the NII safety assessment principles for nuclear power reactors (Ref 1). In Section 3, the design reservations on the RBMK reactor, as expressed in a UK report (Ref 2) of a visit to the USSR in 1975, are given with comments on their relevance to the Chernobyl design.

Legasov, the leader of the USSR delegation to the Vienna conference stated, "The designers made a colossal psychological mistake". They placed too much responsibility on the operators and did not cater for the possibility of the succession of gross violations of practice that occurred.

In acknowledgement of this, the Soviets closed down some of their RBMK reactors to make technical changes. These aim at addressing the problems of the positive void coefficient, the resultant positive power coefficient and the shutdown characteristics. The following measures have been proposed and some are being implemented.

Design changes to the RBMK Reactors proposed by the Soviets

In the information compiled by the Soviet's for the IAEA Experts' Meeting at Vienna (Ref 3) and also in the Summary Report on the Post-Accident Review Meeting on the Chernobyl Accident (Ref 4) the Soviet's gave a list of proposed design changes to their RBMK reactors.

Short term measures to be implemented soon

To prevent development of a control rod configuration which violates the demand for an appropriate 'reactivity reserve', the following steps will be taken:-

1. All rods will be fitted with limit switches ensuring a minimum insertion of 1.2m,

2. and the equivalent of 70-80 rods will be kept within the core.

These two steps will also greatly reduce the value of the void coefficient of reactivity and the overall protection against design basis faults will undoubtedly increase:

3 Violation of the requirement to avoid operating below 700MW(th) will be prevented by additional shutdown protection.

<u>In the longer term the following measures are proposed</u>

4 To mitigate the problem of the positive void coefficient fixed absorbers will be installed and

5 the fuel will be modified by increasing its enrichment from 2% to 2.4% to compensate for the reduction of reactivity caused by (4).

6 Another long term development is a fast acting shutdown system, options for which are being studied.

7 A number of additional indicators of the cavitation margin of the main circulating pumps are being installed,

8 and also a system for automatic calculation of reactivity with an emergency shutdown signal when the reactivity reserve falls below a specified level.

These measures will have an adverse effect on the economic parameters of nuclear power plants with RBMK reactors but they should ensure that a Chernobyl-type accident does not recur.

In addition to the technical measures, organisational steps are being taken to reinforce technological discipline and to improve the quality of operations.

These further measures include:-

9 The quality of training and retraining of the staff needs to be improved.

10 Quality assurance of plant components during manufacture, assembly and construction must be verified more carefully.

11 There will be an expansion of research on quantitative probabilistic analysis of safety.

12 Research on the possibility of building reactors with passive safety systems - so called 'intrinsically safe' reactors, is being expanded.

These technical changes to the design are sufficient, we believe, to avoid recurrence of accidents of the Chernobyl type.

Since this paper was issued in March 1987 the Soviet's have given a further account of the measures they have and are undertaking to enhance the safety of the RBMK reactors (Ref 5).

The proposed short term measures dealing with reactivity reserve, items 1 to 3 have been confirmed by the Soviet's with some modifications. The first measure taken to ensure a minimum control rod insertion in the core of 1.2m, led to a distortion of the vertical field of reactivity and made it necessary to reduce power by 10 to 15%. The present design of the rods has been changed; the connecting link between the rod and the graphite displacer follower has been lengthened. This now enables the control rods to be fully withdrawn without raising the displacer above the bottom of the core so that positive scram can be avoided and it also overcomes the distortion of the vertical flux profile. Fig 1 shows the arrangement of the rods and graphite displacers (a) before the accident (b) after the first modification and (c) after the final modification (Ref 5). (This arrangement is discussed further with respect to 'positive scram' in Section 5).

Furthermore, certain other improvements have been made in the control and protection system which increase the reliability and safety of the reactor operation. These are:-

- The numbers of short absorber rods inserted into the core from beneath has been increased from 24 to 32 on the RBMK-1000. (These rods are used for axial trimming of the power).

- A scheme has been developed for the insertion of short absorber rods into the core in response to emergency protection system signals.

- The digital excess reactivity display now gives readings for any and all states of the reactor, and

- Automatic shutdown of the reactor will now occur when the excess reactivity is reduced to 30 manually operated rods.

For the longer term measures, items 4 to 12, the Soviet's have confirmed that they intend carrying out these measures and also a number of others aimed at improving safety. In particular additional illuminated indicators have been installed on the operators control panel for the reactor's emergency protection system. Another approach to the improvement of safety at existing RBMKs lies in a substantial enlargement of in-core power density control. Modernisation of the existing diagnostic and parameter recording should make it possible to identify and determine the nature of accident situations as they develop, and also to establish the actions of personnel. This improved system is to be removed to a separate complex with its own reliable power supply.

For the reactors at present under construction the Soviet's are considering the possibility of reducing the void coefficient of reactivity by limiting the amount of graphite in the reactor core. They can achieve this by cutting off the fins of the graphite blocks. These measures together with the increased enrichment of the fuel and increased numbers of absorbers should allow the reactor to work with greater safety.

References

1 "Safety assessment principles for nuclear reactors" HM Nuclear Installations Inspectorate, HMSO, 1979.

2 "The Russian graphite moderated channel tube reactor", NPC(R)1275, March 1976.

3 "The accident at the Chernobyl nuclear power plant and its consequences", information compiled for the IAEA Experts Meeting, 25-29 August 1986, Vienna, by the USSR State Committee on the Utilization of Atomic Energy.

4 "INSAG summary report on the post-accident review meeting on the Chernobyl accident", Vienna 30 August - 5 September 1986.

5 "The Chernobyl nuclear power station accident: one year afterwards". IAEA Conference on nuclear power performance and safety, Vienna 28 September to 2 October 1987. IAEA-CN-48/63. 65 pages, Risley Trans. 5541, V G Asmolov and others.

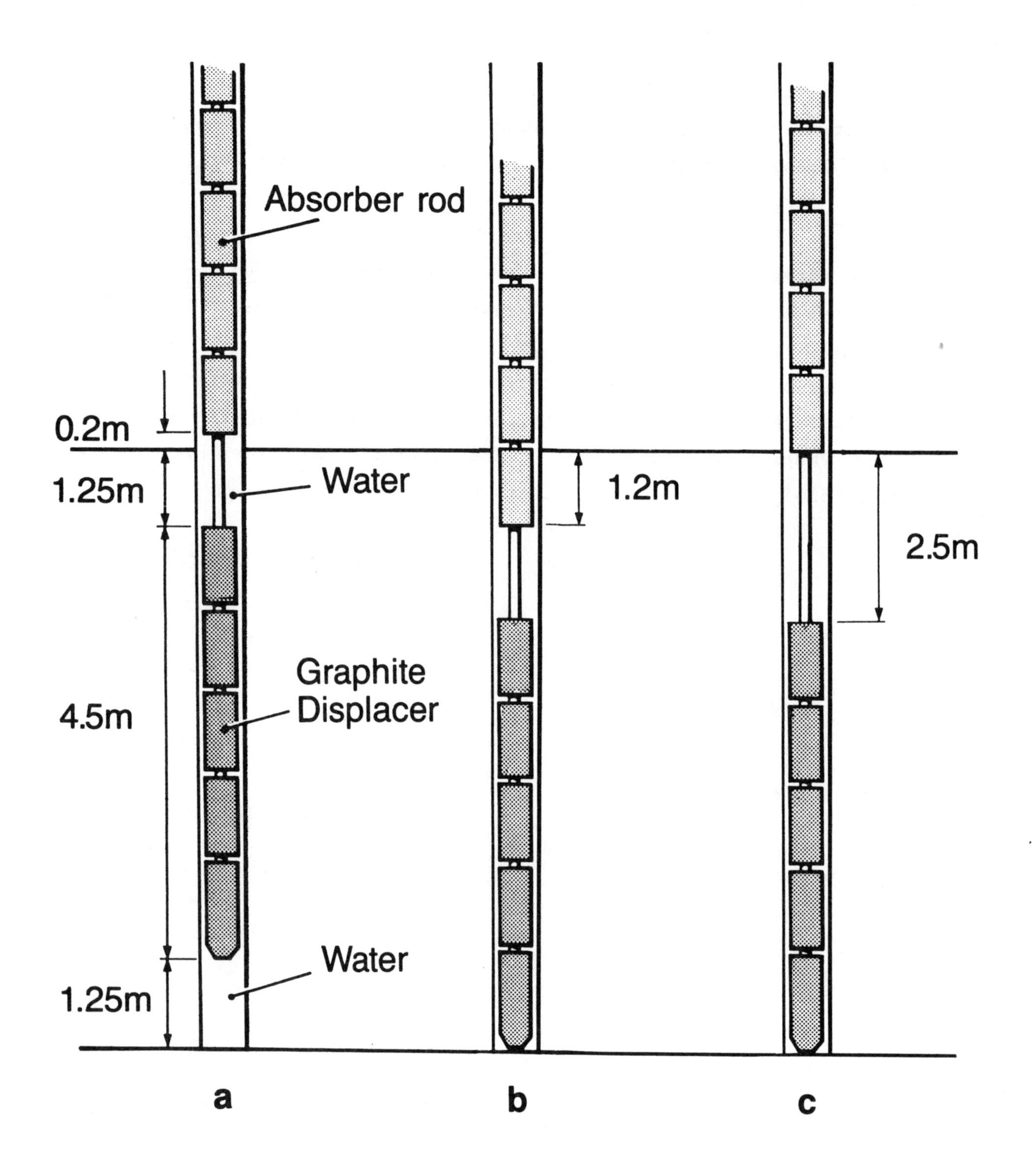

(a) before the accident
(b) after the first modification
(c) after the final modification

FIG. 1 Extreme upper position of control rods

SECTION 5: THE ACCIDENT AT CHERNOBYL

Following the accident at Chernobyl, there was naturally concern about its impact upon the status of nuclear power worldwide. In the early days, it was not easy for Western experts to respond to these concerns because very little information about the accident was available. However, under the leadership of Dr Blix, the IAEA organised an international conference in Vienna in August 1986, at which Soviet scientists and engineers made comprehensive presentations on the accident.

The Chernobyl accident was an event with considerable implications for both the Soviets and the rest of the world. Nonetheless, although the responsibility for the decisions and events that led up to the accident rests with the Soviets, they command respect for the strenuous efforts they made to control the accident once it had occurred. All who are professionals in the nuclear business have been impressed with the bravery and the dedication of the operating engineers, firemen and men on the spot once the accident occurred. The Soviet authorities also moved with incredible speed and efficiency to control the accident, mitigate its effects and command the recovery process. Speaking in Geneva after the accident, Lord Marshall said "judging from present information, we shall be full of praise for the recovery operation itself." Now that the Soviet report (Ref 1) on the accident has been published, that point is reiterated.

In addition, the international community must be grateful to the Soviet authorities for the frankness with which they have described this event. All who went to the Conference in Vienna came away with the strong conviction that the Soviet engineers and technical experts had told everything that they knew themselves about the accident. There are still some points of detail to be resolved but there is general satisfaction that it is now known why the accident happened, how it happened and how it should have been prevented.

As a result of this, the implications of the Chernobyl accident for the UK and the question "could Chernobyl happen here?" can now be addressed. On consideration, it seems certain that a Chernobyl-type accident could not happen in the UK. UK safety rules first and foremost aim at the building of reactors that have intrinsic characteristics that provide inherent protection. Secondly, these natural defences are supplemented by engineered features to prevent, limit, terminate and mitigate any faults. Thirdly, the systems design must be tolerant to operator action - if the operator makes a mistake, the reactor shuts down. Fourthly, UK operators are highly educated and well trained, not just for routine operations but for unusual situations and accident situations and, fifth, the entire system is overseen by an independent Nuclear Inspectorate that can at any time, without hindrance or challenge, close down any licensed reactor.

The reasons for the accident at Chernobyl are now clear. It occurred as a result of three main design drawbacks of the reactor:

1 it had a positive void coefficient and, below 20% power, a positive power coefficient, which made it intrinsically unstable at low power;

2 the shutdown system was in the event inadequate and might in fact have exacerbated the accident rather than terminated it;

3 there were no physical controls to prevent the staff from operating the reactor in its unstable regime or with safeguard systems seriously disabled or degraded.

Malpractice by the operators was a contributory cause and indeed, they broke so many rules that one cannot help thinking that this was their regular habit. It is not credible that this was the one and only occasion on which they behaved in this manner. The Soviets now acknowledge that the reactor designers should have envisaged such possibilities and guarded against them. However, they also pointed out that the RBMK reactor had been designed before microprocessor controlled protection systems had been developed and at that time they placed more confidence in proper operator action than in automatic safety circuits, which they considered then to be less reliable.

Gross maloperation of the reactor resulted in the core containing water at just below the boiling point, but little steam at the start of a turbogenerator experiment. When the experiment began, half of the main coolant pumps slowed down and the flow reduction caused the water in the core to start boiling vigorously. The bubbles of steam that formed absorbed neutrons much less strongly than the water they displaced and the number of neutrons in the core started rising. Technically, such behaviour was a consequence of the "positive void coefficient of reactivity" of RBMK reactors. It was normally counteracted by the "fuel (Doppler) effect", which causes the number of neutrons in a reactor core to be reduced as the fuel temperature rises. At the start of the turbogenerator experiment at Chernobyl however, the conditions were such that the void effect was dominant and the "power coefficient" of the reactor was positive. This was a dangerous condition, for it meant that if a disturbance was to cause the power to rise, the number of neutrons in the core would also rise, amplifying the power increase and causing the power to escalate through a "positive feedback" phenomenon. The Soviets were well aware that under certain conditions, their RBMK reactors were inherently unstable due to this phenomenon, but they relied solely upon the operators obeying regulations to prevent operation under these conditions. These regulations were violated during the turbogenerator experiment at Chernobyl and the slowing down of the coolant pumps provided the disturbance that caused the reactor power to run away.

As the number of neutrons and power of the reactor rose, more steam was produced and even fewer neutrons were absorbed thereby increasing the "excess reactivity" of the reactor. The operator realised that the reactor power was rising and pressed the emergency shutdown button. Not only was the shutdown system too slow in its operation, however, it also appears to have had a serious design flaw (positive scram) that in the event could cause it to increase significantly the magnitude of the developing power surge (Ref 2). The excess reactivity rose to the level where the "chain reaction" could be sustained by "prompt neutrons" alone and the reactor became "prompt critical" before the shutdown system could become effective. The power surge caused the fuel to heat-up, melt and disintegrate. Fragments of fuel were ejected into the surrounding water, causing steam explosions that ruptured fuel channels and led to the pile cap being blown off.

Several features of RBMK operation played a role in the development of the accident at Chernobyl and so in this Section the normal operation of the reactor is briefly outlined first to provide a point of reference. The development of the accident is then described and discussed. In probabilistic safety assessments, nuclear reactor accident sequences are usually broken down into a number of steps such as: accident initiation; cessation of fission by reactor shutdown; provision of cooling to avoid core degradation; and containment of fission products released from the pressure circuit. It is recognised however, that some initiating events might be so severe that the engineered safeguards of reactor shutdown and core cooling would be unable to cope with the conditions arising, although the reactor design should make such events highly improbable. The steps listed above would not then provide an appropriate framework for the description of the accident. The initiating event at Chernobyl fell into the latter class, so rather than consider the accident in terms of the steps above, it will be discussed under the headings: accident initiation; failure of shutdown system; core degradation and pressure circuit failure; and response to the accident.

That the design of the RBMK reactor is deficient has already been made abundantly clear in this Report. Thus Section 2 revealed that the reactor goes against many of the most important of the safety principles enunciated by the UK Nuclear Installations Inspectorate and Section 3 described the safety reservations that the NNC had when they assessed RBMK reactors ten years ago. Section 4 shows that the Soviets now propose to make significant changes to the reactor in order to render such accidents less likely in future. In this Section, too, reference will be made where appropriate to the NII safety principles (Ref 3) and the CEGB design safety criteria (Ref 4), showing how contravention of these led to the various phases of the accident.

5.1 Normal operation of a reactor

Most power stations generate electricity via a thermal process, that is heat is used to boil water and raise steam that is fed to turbogenerators to generate electricity. In nuclear power stations, the heat source is a nuclear reactor in which fission of the nuclei of atoms of certain heavy elements (known as fissile elements) is induced by collisions with neutrons.

Fission of a nucleus releases a relatively large amount of energy (about 200 MeV), which is divided amongst the products of the fission. These are principally two large fragments of the nucleus, which are themselves nuclei, plus neutrons and gamma rays. Over 80% of the energy released manifests itself as kinetic energy of the fission fragments and this is rapidly converted to heat in the fuel as the fragments collide and slow down. Additional heat is produced by the absorption of neutrons and gamma rays in the reactor, and by radioactive decay of fission products. Neutrons released in fissions are themselves responsible for inducing further fissions, and so a "chain reaction" is established in which fissions at one time provide the conditions for further fissions at a subsequent time.

In a steady state, the rate of fission at any given time must be equal to that at any preceding time, which in practice means that the number of neutrons in the reactor must be the same from one generation to the next. The reactor is then said to be "critical". On average however, about $2\frac{1}{2}$ neutrons are emitted per fission, so just under half of the neutrons produced must induce fissions in the reactor. The remainder are either absorbed within the reactor, or escape from it.

All the components of the core of a reactor absorb neutrons. Only a small proportion (about 2%, depending on the fuel enrichment) of the heavy elements in the fuel are fissile and some neutrons are absorbed in the fuel without causing fissions. These are mainly captured by the isotope of uranium, ^{238}U, which is a major constituent of the fuel. Some of these captures result in the formation of the plutonium isotope ^{239}Pu, which is itself fissile and so the depletion of the initial charge of fissile elements is to some extent compensated by the production of new fissile material during reactor operation. Furthermore, capture by ^{238}U of neutrons with energies less than about 1 KeV results in an effect with an important bearing on reactor stability. So called "resonances" between low energy neutrons and ^{238}U nuclei cause efficient capture of neutrons whose energies lie within certain very narrow bands. As the fuel temperature rises, the widths of these bands increase due to thermal motion of the absorbing nuclei, and this increases the probability of neutron capture. Consequently if a small disturbance were to raise the power of a reactor, the fuel temperature would rise, neutron capture in the fuel would increase and so the rate of fission in the reactor would tend to decrease. A reduction of the rate of fission would reduce the reactor power and so this neutron capture effect would tend to stabilise the reactor against disturbances in power. This effect is expressed technically through the Doppler coefficient of reactivity and it is an important example of an intrinsic or inherent safety feature of the reactor.

Neutrons are also absorbed in structural materials, the coolant, moderator and control rods. Except for absorption in control rods, such absorption is disadvantageous and materials are chosen to minimise neutron absorption, subject to other design requirements. At Chernobyl, light water was used as coolant and graphite as the moderator. In some reactors eg pressurized water reactors, light water plays the dual role of both coolant and moderator. Although a moderator absorbs neutrons, its principal function in a reactor is to minimise non-productive neutron capture. It does this by scattering the high energy neutrons produced by fission so that their energies are rapidly reduced to values below those at which resonant capture in the fuel could occur. This reduces non-productive capture in the fuel, allowing a chain reaction to be sustained in fuel with low fissile material content. The combination of water coolant and graphite moderator in the Chernobyl reactor was disadvantageous for reactor stability however, since the graphite alone could effectively moderate the nuclear reaction and the main neutronics effect of the water was therefore as an absorber. The designers of the RBMK reactor could have avoided this undesirable characteristic, but they chose instead to accept it in order to optimise the fuel burn-up and economics of the reactor. Consequently an increase of the volume of steam in a fuel channel, expressed through the void fraction, would reduce neutron absorption in the channel and tend therefore to increase the reactivity and power of the reactor. This is expressed technically through the positive void coefficient of reactivity. Since a small disturbance increasing the power would increase the void fraction in fuel channels and this would tend to increase the power further, the positive void coefficient exerts a destabilising influence on the reactor. Operating above 20% of full power, the stabilizing effect of the Doppler coefficient dominates the destabilizing effect of the void coefficient in RBMK designs and the reactor could therefore be operated stably. Below this level however, the reactor could enter an unstable regime. This was recognised by the designers and so continuous operation below 20% full power was forbidden (Ref 5).

COMMENT

The main neutronics effect of the water in RBMK designs was as an absorber, so the removal of water from the core caused its reactivity to rise. Being a volatile liquid, the water was furthermore an absorber that could be rapidly displaced from the core by vapour (steam) production. RBMK cores could therefore suffer rapid surges of reactivity as a result of coolant voiding, which could adversely affect the balance between the reactor power and heat removal by the coolant.

For any proposed nuclear reactor in the UK, HM Nuclear Installations Inspectorate requires that specific consideration be given in the design to the effect of coolant voiding on reactivity. The NII Safety Assessment Principles state (Ref 3):

"93. Possible effects of changes in coolant condition on the nuclear reactivity of the reactor core should be identified in the safety submission. Adequate provision should be made to limit the consequences of any adverse change of this kind either by the provision of appropriate protection systems or by the selection of appropriate reactor core design parameters".

The NII requires reactor designs to conform with its Safety Assessment Principles "as far as is reasonably practicable". In respect of Principle 93, the RBMK design clearly does not do so. The Soviets now intend to rectify this design deficiency by increasing the fuel enrichment and number of absorber rods in RBMK reactors. They also intend to add a fast shutdown device - in NII terms an "appropriate protection system".

The control rods provide control of reactor power by allowing neutron absorption in the core to be varied. This is necessary for reactor start-up and shutdown and to compensate for local and core-wide variations of reactivity during operation. The control rods are able to perform this function by virtue of the existence of both "prompt" and "delayed" neutrons in the chain reaction. About 99.5% of the neutrons released in nuclear fission are released essentially as soon as fission occurs and are known as prompt neutrons. They slow down and diffuse in the graphite moderator and are absorbed causing further fissions in a few milliseconds. If a reactor were critical on prompt neutrons alone, changes of power would occur on a timescale much shorter than the characteristic response timescale of any power control system and reactor control would be impossible. Fortunately, the 0.5% of neutrons that are not emitted promptly are emitted after delays of up to about one minute and provided that a reactor is operated so that is is critical only when both prompt and delayed neutrons are accounted for, reactor control presents no difficulties. Reactors are therefore designed to avoid prompt criticality or limit its consequences.

COMMENT

The control rod system on RBMK reactors before the Chernobyl accident had a serious design shortcoming. The inclusion of part-length control rod followers, intended to increase control rod effectiveness and minimise fuel cycle costs, allowed the reactors to be manoeuvred into a situation where the insertion of control rods increased reactivity rather than reduced it (see Section 5.2c and Figure 4 for details).

In a presentation to the IAEA in September 1987 (Ref 9), the Soviets estimated the magnitude of this "positive scram" effect in the Chernobyl accident to be in the range 0 to 1.5β based upon reactor measurements made just before the accident, where β is the delayed neutron fraction. A step increase of reactivity of 1β would make the reactor prompt critical.

The possibility of such an effect would not be permitted in UK reactors since it would clearly violate the NII Safety Assessment Principle 143:

"143. ... the shutdown system should be capable of shutting down the reactor and holding it sub-critical with a margin of negative reactivity which should be available at all specified times and which should allow for uncertainties in nuclear characteristics, perturbations in plant state etc.."

5.2 Accident initiation

At Chernobyl, an initiating event too severe for the engineered safety systems and containment occurred. The design of the reactor core itself made such an event possible, but in recognition of this, operating instructions had been written to prevent the relevant conditions arising. Unfortunately, these operating instructions were violated, opening up the possibility of the severe initiating event. The performance of an experiment then produced the necessary trigger for the event to occur.

Ironically, the experiment that triggered the initiating event was designed to improve the safety of the plant. The objective was to see whether the mechanical inertia in a turbogenerator, isolated from both its steam supply and the grid, could be used to supply electricity to essential systems for a short period following a power failure. In essence what was being attempted was to use the turbogenerator as a mechanical "flywheel" coupled to the main reactor circulation pumps electrically.

When disconnected from the grid and steam supply, a free-wheeling turbo-generator would take about 15 minutes to come to rest from 3000 rpm but when coupled to the pump motors might provide a few tens-of-seconds supply. Even so, given the rapid coast down of the pumps without this provision, the long 'scram' and diesel start times, this "flywheel" effect would have provided a valuable margin in the safety case.

The experiment had been attempted twice before in 1982 and 1984. The 1982 experiment could not have involved Chernobyl reactor number 4 since it had only been in operation for two years when the accident occurred. In 1984, following isolation of the generator from the grid the voltage level in the unit system fell rapidly and the operators were unable to arrest the drop by manual control of the voltage regulator. The fall in voltage resulted in the pump motors slowing down much faster than the generator. Nonetheless, although unsuccessful on the previous two occasions, the experiment had not then led to disaster. This was because the reactor had presumably not been manoeuvred into the same dangerous condition as Chernobyl Unit 4 prior to the experiments and because the reactor had been tripped at the start of the previous experiments.

For the experiment on the 26 April an automatic voltage regulator acting on the generator excitation current had been fitted with the aim of maintaining the voltage level so that the pump motors ran down in step with the main generator at synchronous speed, drawing upon its stored kinetic energy.

The planned experimental initial conditions required the reactor to be at about 25% full power with one of its turbogenerators shut down. The other turbogenerator would supply two main circulating pumps in each loop. The remaining two pumps in each loop and the auxiliary plant were to be fed from the grid.

5.2a Chronology of the accident sequence

25 April 1986

01.00 Commencement of power reduction for maintenance shutdown.

13.05 50% power level (1600 MW(th)) achieved. Turbogenerator number 7 was disconnected from the grid and all house load was transferred to the still operating number 8 unit.

14.00 In accordance with the programme for the turbogenerator experiment, the reactor's ECCS was disconnected. Power was then held at 50% as the start of the experiment was delayed by a request from the controller in Kiev to keep supplying electricity to the grid. The ECCS was not switched back on, in violation of the operating rules.

COMMENT

The Soviets recognise this as the first fault (Ref 1). The disconnection of the ECCS was part of the approved experimental programme and had little effect on the course of the accident. Nonetheless, had a demand for the ECCS arisen during the period of operation at reduced power, it would not have been immediately available and a further route to core damage would have been opened up. The risk of an accident was thus raised unnecessarily by continuing operation without the ECCS connected.

The failure to reconnect the ECCS when the experiment was delayed violates a 'fundamental safety principle' of the NII and CEGB:

"All reasonably practicable steps shall be taken to prevent accidents (Ref 3, NII principle 4; Ref 4, Foreword).

It also violates two of the NII's engineering principles:

"100. Provision should be made for a sufficient and reliable supply of reserve coolant, separate from the normal supply, to be available in an adequate time in the event of any significant leakage of primary coolant";

"107. Adequate protective systems should be provided and, whenever fuel is in the reactor, they should be maintained at a level of readiness adequate to ensure nuclear safety".

23.10 The power reduction programme was resumed. The aim was to perform the test with the reactor at between 700 and 1000MW. On going to lower power, that set of reactor control rods used to control the power of the reactor at high powers (called the Local Automatic Rods), was switched out at 00.28.00 hrs and a set of rods called the Automatic Rods switched in. However, the latter were set to continue power reduction rather than hold the power level as required and the operator was unable to stop the power of the reactor falling to 30MW(th).

COMMENT **This is the second fault recognised by the Soviets. The reactor became difficult to control. Administrative rules prohibited power operation (except start-up and shutdown) below 20% of full power because of the difficulty of controlling the thermal-hydraulic and reactor physics conditions in the core below that power. Accordingly, the experimental programme required that the reactor power be above the 20% limit for power operation. Because of an error during the changeover of controlling systems, the power level fell below 20% and the experiment should have been aborted. Neither basic design nor additional automatic safeguards prevented the reactor from continuing to operate in this inherently unsafe regime.**

A design of reactor core that permitted operation under conditions of short-term instability would not be acceptable in the UK, for it would be difficult to demonstrate conformity with the NII principle:

"68. Where changes of condition or state of components within the core, such as temperature changes or coolant voiding, can adversely affect core reactivity, precautions should be taken in design and operation to avoid or minimise the effect

of such changes by the use of adequate design margins and limitations of operating conditions".

26 April

01.00 Operator succeeded in stabilizing the reactor at 200MW(th). The combination of low coolant voidage at low power, the positive void coefficient and xenon poisoning made it difficult to achieve this level and the operator did so only by removing further control rods from the core. His reactivity reserve (ie potential increase of reactivity if all control rods were withdrawn) at this time was well below the limits laid down in the regulations.

COMMENT **The third fault listed by the Soviets. The reactor design required the operators to position the control rods in the core in such a manner as to ensure a sufficiently rapid shutdown capability and to minimise positive reactivity effects. This was expressed as a requirement to have a reactivity reserve no less than a certain minimum value, given by an equivalent number of fully inserted control rods. This requirement was only achieved, however, by an administrative operating rule and not by any engineered interlock. The reactor was thus able to continue low power operation.**

It is remarkable that the design relied upon a single system for reactor shutdown and that this system relied on the observance of an administrative rule to maintain its effectiveness. Clearly, a single failure within the protection system, be it either mechanical or administrative, could prevent timely and effective reactor shutdown. This would be unacceptable in the UK.

From the UK viewpoint, the RBMK's emergency protection (shutdown) system has several major shortcomings associated with its reliance upon a

single system to achieve shutdown. First, it does not conform to one of the NII's general engineering principles for nuclear plant design, ie:
"31. The plant should be designed and operated in such a manner that no single failure should lead to a radioactive release or the occurrence of any direct radiation in excess of the requirements of principles 13 to 17".

This general principle is reinforced by specific engineering principles for the protection system. The RBMK design does not conform to several of these:

"112. No single failure within the protection system should prevent any protective action achieving its required performance in the presence of any specified fault or external hazard initiating a demand on the protection system";

"115. The required performance of components, subsystems and systems should be stated and shown to be adequate for the purpose of providing protection. Limits should be defined outside which components etc should not be operated and provision should be made to ensure that these limits are not infringed. It should be shown that the overall reliability of the protective system is adequate";

"121. It must be recognised that unforeseen plant or protection system faults or maloperations may occur. Protection system design should reflect this aspect by, for example, the provision of reasonably practicable diversity and redundancy, both within each system and in the nature of each input and output";

"122. Diversity of fault detection and protection should be employed where reasonably practicable but where protection system reliability is required to be very high or when there is doubt about the reliability or effectiveness of a non-diverse system diversity should be introduced".

These principles are reflected in the CEGB's design safety criteria, which require that (Ref 4, sub-section 5.3):

"wherever practicable, systems shall be designed to be fail-safe in the event that a fault occurs in that system.

For those faults which originate in the control or protection equipment and which could lead to a release, no claim shall be made for the control or protection function provided by this equipment. Alternative protection systems of appropriate reliability shall be provided to ensure reactor safety under such conditions."

As explained in Section 4 of this Report, the Soviets have belatedly recognised these short-comings and intend to install rapid-acting diverse shutdown systems on RBMK reactors in the future (Ref 5). The existing control rod system has been modified to ensure that the lower ends of the control rod followers cannot rise above the bottom of the core, thus preventing any possibility of "positive scram".

01.03 and 01.07 One additional main circulation pump was switched into each coolant circuit. This made a total of 8 working. Following the experiment, four pumps would remain in operation, four having run down during the experiment.

Switching in these pumps increased the flow rate into the core. Since the reactor was already at low power, the hydrodynamic resistance was very low and the flow rate of water through the core was very high. Some pumps were operating beyond their permitted operating regimes. The increased flow caused a reduction in steam formation and a consequent fall of pressure and water level in the steam drums. By this time the voidage in the water being circulated through the reactor was tiny. The coolant subcooling in the pressure circuit was reduced and the whole circuit brought close to boiling.

COMMENT

Fault number 4. Pump operation beyond permitted operating regimes could have led to pump damage. The startup of the additional pumps might have been part of the experimental programme, but as the reactor power was much less than that specified in the programme, the thermal-hydraulic conditions were very different to those anticipated. The result was that steam formation in the core was greatly reduced, lowering the core reactivity but providing the starting point for a large potential change in coolant voidage.

01.19 Operators then tried to increase the water level in the steam drums by using the feedwater pumps - the reactor should have shut down on low water level in the steam drums but the operators had been able to disengage that signal. The cooler feedwater reduced steam generation further, the reactivity continued to drop and the operator compensated for this by removing manual control rods from the core. This reduced the reactivity reserve still further.

COMMENT

The fifth fault noted by the Soviets. The reactor protection system was disabled. Difficulties in reactor control and the potential for a reactor protection trip had led the operators to override the protection signals from the steam drums. Again the actions illustrate deficiencies in design that

would be unacceptable in the UK. UK reactors must be designed so that access to protection systems is physically restricted to ensure the availability of a specified minimum amount of operational protection equipment.

This requirement is embodied in the NII safety assessment principles that state:

"38. Unauthorised access to and interference with safety-related structures, systems and components should be prevented by suitable means;"

"127. The minimum amount of operational protection equipment for which reactor operation will be permitted should be specified...;"

"131. The design should be such that the means of access to all protection equipment can be physically controlled to limit access to an extent which ensures availability of the minimum amount of operational equipment referred to in principle 127."

UK reactor designs conform to these principles by incorporating lock-and-key systems and engineered interlocks into the reactor protection systems. Such features were not, however, to be found on RBMK reactors.

01.19.58 Turbogenerator bypass valve closed. Steam was no longer being dumped in the condenser and the fall of pressure caused by the feedwater flow lessened.

01.22 Operator reduced feedwater flow. The supply of relatively cool feedwater to the reactor was substantially reduced.

COMMENT

The thermal-hydraulic conditions in the core were now such that a rapid and large change of coolant voidage could occur. The operator did not, however, appreciate the state of the plant and the information and data available to him probably did not provide the basis for such a diagnosis.

The importance of providing plant operators with clear and comprehensive information on plant state is well recognised in UK nuclear plant designs, which provide instrumentation to alert operators to all conditions pertinent to plant safety and alarms to signal when set limits are reached.

This importance is reflected in the NII safety assessment principles:

"66. The core should be designed such that all safety-related conditions can be monitored to an adequate degree of accuracy;"

"132. Provision should be made in the form of indicating and recording instruments to inform the plant operators at all specified times of the state of those items which have a significant influence on safety and on safety-related aspects of the overall plant state. Such provisions should include devices to give advance warning of unacceptable changes and rates of change and also alarms when set limits are reached. Sufficient information should be made available to the operator at all times to enable an accurate appreciation to be made of the plant state so that all actions necessary in the interest of safety can be taken promptly and effectively....;"

The Soviets have recognised design deficiencies of RBMK reactors in this area. They intend to install

additional instrumentation to monitor the cavitation margin at the main coolant pump inlets.

01.22.30 A print-out of the reactivity reserve was produced - it was only 6-8 rods and immediate shut down was required. However there was no automatic shutdown on this signal and so the reactor remained critical.

COMMENT **The reactor design is here again shown to be deficient. It should have embodied devices which <u>automatically</u> shut down the chain reaction under such dangerous circumstances. Instead reliance was placed on the interpretation, by the operator, of a computer print-out.**

The computer print-out was requested by the operator, probably so that the initial conditions for the turbogenerator experiment would be known. Amongst the data on the print-out was information about control rod positions that showed the plant was well outside the permitted operating regime. No alarm was given to warn of this unsafe state. If the operator saw this information, he would have recognised the message. It appears that he was either unaware of the information or did not appreciate the severity of the plant state and the potential consequences for plant safety.

The NII safety principles require that UK nuclear reactor designs are also safeguarded against such an ergonomic deficiency, eg

"129. Alarms should be provided to give warning that any safety-related system, component or parameter is at a pre-set limit of its acceptable operational state. Where reasonably practicable

alarms should be initiated in the event of any unsafe failure of any element of a protective system."

01.23.04 Experiment began; the regulating valves to turbogenerator number 8 were closed. Reactor power was still about 200MW(th). The shutdown signal for loss of two turbogenerators had been blocked by the operators to permit a re-run of the experiment if the first were not successful.

COMMENT **Sixth and final fault. Once again, the design of the reactor had allowed the operators to disable a protection system, demonstrating yet again that it would not conform to UK safety standards as embodied in the NII's safety assessment principles numbers 38 and 131. It is not clear, however, whether a trip of the shutdown system at this stage would have prevented the accident. The distorted axial neutron flux shape and the position of the control rods (almost all fully out) might have led to a transient increase in reactivity (positive scram), merely causing the accident to happen 36s earlier.**

The power of the reactor began to rise slowly.

01.23.40 The operator ordered full emergency shutdown. Three seconds later, there were high power and short period alarms. All control and shutdown rods were being motored into the core but they were too late. Not all rods reached their low stops, the operator uncoupled rods to fall "under their own weight". Shocks were felt.

COMMENT **It is noted that as a result of disabling automatic protection systems, the operator had to order manual shutdown of the reactor. The capability to negate correct protection system action would be unacceptable**

in the UK.

This shows yet another lack of conformity with the NII's safety assessment principles. Principle number 124 states:

"124. The protection system should be automatically initiated. No operator action should be necessary in a timescale of approximately 30 minutes. The design should however be such that an operator can initiate protection system functions and can perform necessary actions to deal with circumstances which might prejudice the maintenance of the plant in a safe state but cannot negate correct protection system action at any time."

This is reiterated by the CEGB's design safety criteria (Ref 4, subsection 5.3.5).

Despite this, a rapid-acting shut down system might have prevented prompt criticality and the Soviets say they will fit such a system (see Section 4).

01.24 Observers outside the reactor at about this time reported two explosions about 3-4 seconds apart. Burning lumps of material and sparks were thrown into the air - some landed on the turbine hall and started fires. There was a danger that the fires would spread along the turbine hall roof to the adjoining Unit 3, so the immediate response of the firefighters was to extinguish the fire on the roof. It was during this action that many of the firefighters received fatal doses of radiation.

COMMENT **In the UK, the effect of plant layout and building materials on the potential for accident propagation is considered as thoroughly as other aspects of nuclear plant safety and designers are required to**

minimise this potential. In particular, non-combustible materials must be used wherever practicable and failing that, steps must be taken in the design to reduce the rate of spreading of fire.

The NII safety assessment principles include consideration of plant layout, as for example in:

"258. The layout of the reactor and other safety related plant should be such as to minimise the effects of external hazards and of any interactions between a failed structure, system or component and other safety-related structures, systems or components."

The CEGB goes further still by requiring (Ref 4) that "non-combustible and heat-resistant materials shall be used wherever practicable throughout," and "where it is impracticable to use non-combustible materials, then installations and materials specifically designed to reduce the rate of propagation of fire shall be used."

5.2b Information derived by the Soviets

RBMK reactors are equipped with the so-called "Skala" centralised control system, which periodically records conditions in the reactor. For the turbogenerator experiment, only those conditions important for the analysis of the experimental results were recorded at high frequency. Nonetheless, data from the "Skala" system, combined with instrument readings and reports from the reactor operators, has enabled the Soviets to validate a computer model of the initiation of the accident. The results of their analysis were shown in Figure 4 of the Soviet report (Ref 1), which is reproduced here as Figures 1a-1c.

A US Department of Energy (USDOE) team has put a considerable effort (Ref 6) into obtaining a full understanding of this figure. As a result, they have been able to draw attention to two areas where particular care is needed in its interpretation. First, the control rod motion in the figure is hypothetical, although it is clear that the local automatic control rods were being inserted during the final moments of the accident. Second, the USDOE analyses have shown consistently that the pressure build up in a closed system suppresses a power excursion, so steam must have been released. It is understood that the Soviets have since confirmed that this did indeed occur. This lends support to a USDOE claim that curves M and O of the Soviet Figure 4 are labelled incorrectly. The USDOE team believe that curve M does not represent the total steam leaving the system as it indicates that no steam was released during the development of the power surge. They also believe that curve O represents the total steam flowrate, on a scale of 0 to 600 kg s^{-1}, rather than the steam mass quality at the core exit as indicated in the Soviet report.

In normal operation, the pressure difference between the core inlet headers and the steam drums was about 1.2 MPa and the temperature of the coolant at the inlet was about 14C below the boiling point. At the start of the test however, the steam content in the pressure tubes was tiny and their hydraulic resistance was low, reducing the pressure difference between inlet headers and steam drums to an unusually low value. Furthermore, the heat being generated in the core (200 MW) was only just sufficient to supply the heat of vaporisation required by the steam flowrate (about 80 kg s^{-1}, using the USDOE interpretation of curve O). Consequently the coolant temperature at the core inlet was only about 1C below the boiling point. The coolant conditions were therefore such that a large, rapid change of voidage could occur, which would cause a large increase in core reactivity via the positive void coefficient.

COMMENT

A reactor design that allowed such a circumstance to arise should not be acceptable in the UK, for the reactivity could rise rapidly, leading to a power surge and fuel damage.

It would not conform with the NII's safety assess-

ment principle 94, which states:

"94. ... any adverse change in heat transport or coolant condition which might lead to an unsafe condition should be safeguarded against." The safeguards should, the other principles make clear, be automatic trips with interlocks.

When the experiment started and four of the coolant pumps began to run down, the rate of steam generation increased, causing an initial increase of pressure that reduced voidage in the coolant. This reduced the reactivity and the automatic control rods withdrew to compensate. As the coolant flow reduced further however, steam generation caused the voidage to increase, increasing the hydraulic resistance of the core and tending to accelerate the reduction of flow. The reactivity was increased and the automatic control rods were inserted to compensate. At 1:23:31, these rods were too far inserted to be effective and failed to compensate for the loss of water from the core. The reactor power then started increasing, raising fuel temperatures and heat transfer to water. This caused the coolant voidage to increase rapidly and runaway began. By the time the manual trip of the reactor was initiated at 1:23:40, the action of inserting the rods could have caused a transient increase in reactivity of up to 1.5β, according to Soviet analyses. There were high power and short period alarms at 1:23:43 and within a second, the reactor became prompt critical and the power rose to a value estimated by Soviet analysts to be about 100 times normal full power. The fuel then fragmented, resulting in a rapid increase of heat transfer to water and an abrupt growth of voidage. This caused the flow into the core to cease and the Soviet analysis suggests that a further surge of power to about 440 times normal full power occurred and the fuel channels ruptured. The second power surge depends on the validity of the core representation in the Soviet model however, which would be in some doubt following the first power surge. The Soviets estimate that in the first surge, about 1.3 kJ/g of energy was deposited in the fuel on average, which raised its temperature to about 3000C.

5.2c Discussion of the accident initiation

Western reviews of the Soviet report have all confirmed the plausibility of the Soviet description of events. The initiating event was a reactivity excursion. This resulted from a reactor design that allowed sustained power operation in an unstable regime. Furthermore, it could have been exacerbated by a serious flaw in the design of the RBMK shutdown system causing a transient increase in reactivity when activated.

The void coefficient of RBMK reactors becomes more positive as absorbers are withdrawn from the core. At the time of the turbogenerator experiment, most rods had been withdrawn to counter the reduction of reactivity resulting from having too much water in the core. The void coefficient was therefore positive and unusually large at the time of the test. As there was little steam in the core, the potential existed for a large increase of voidage and reactivity. Furthermore, the thermal-hydraulic conditions were such that a small increase in reactor power resulted in a larger increase of voidage, and hence reactivity, than usual. The void effect dominated the Doppler effect and made the power coefficient of the reactor positive. When the pumps ran down and coolant flow into the core reduced, the rate of boil-off rose, increasing the volume of steam in the core. This tended to increase the reactivity of the core due to the positive void coefficient, but the automatic control rods initially compensated for this effect. When they no longer did so, the power of the reactor started to rise. The positive power coefficient then caused the reactivity to rise still further and the power escalated. Finally as most absorber rods were withdrawn from the core, the speed with which they could effect shutdown on demand was significantly reduced and for the first few seconds of movement they could have increased reactivity rather than reduced it.

An independent order of magnitude estimate of the rise in fuel temperature due to coolant voiding can easily be made. The void coefficient during the power surge could have been typically 25mN/% void, so increasing the void fraction from its initial value of almost zero to about 80% (the normal value at the fuel channel outlets) would have raised the reactivity of the core by about 2 Niles (about 4 times the prompt critical value). To compensate for this, the fuel temperature would have to rise, removing reactivity by the Doppler effect. As the Doppler coefficient would be typically -0.7mN/K at the higher than normal temperatures expected from the power surge, the temperature would have to rise by (2/0.0007 = 2857 K) to overcome the increased reactivity caused by voiding. The fuel temperature would therefore rise to about 3150C. This, and the Soviet value, are supported by more sophisticated calculations performed by the CEGB using the RELAP computer code.

This essentially qualitative description of events is supported by calculations performed by a USDOE team (Ref 6). They have analysed the Chernobyl power surge by using the MINET and CRAS thermal-hydraulics computer codes, both of which include a simple reactivity model. These code runs clearly demonstrated the type of transient displayed in the Soviet Figure 4, that is a power excursion driven by the large positive void coefficient (see Figure 3). They also showed that steam must have been released from the system during the transient, for otherwise the power surge would have been suppressed. The reason for this is that the compressibility of water is low, so steam-filled voids could only have grown significantly had coolant release from the pressure circuit occurred to allow them to expand. With both codes, however, it was necessary to introduce a transient positive reactivity addition (10c for MINET, 1$ for CRAS) at about the time that the emergency shutdown button was pressed in order to obtain detailed agreement with the timing of the power surge in the Soviet Figure 4. The USDOE team noted several possible mechanisms for such a reactivity addition and placed emphasis on a potentially serious flaw in the design of the shutdown system.

The emergency shutdown signal caused all the top-entry control rods to be motored into the core. The control rods on the Chernobyl reactor moved within dedicated control rod channels and were cooled by water in a closed circuit separate from the pressure circuit. In RBMK reactors, however, the main neutronics effect of water is as an absorber so if water took the place of the control rods as they were withdrawn from the core, the effect of withdrawal would be reduced. To counteract this, long non-absorbing graphite rods known as followers were suspended below all of the top entry control rods, to take their place as they were withdrawn (see Figure 4). Due to height restrictions within the reactor however, the graphite followers had to be made shorter than the fuelled region of the core.

Consequently, when the control rods were fully withdrawn from the core, the followers did not occupy the full length of the control rod channels and the bottom 1.25m contained just water. A potentially dangerous situation could therefore arise. If control rods were inserted from the fully withdrawn position, the water at the bottom of the core would be displaced by the graphite followers and there could be a transient local positive reactivity addition, the so-called "positive scram" effect. Such was the situation after the operators of the Chernobyl reactor noticed that the power was beginning to rise and pressed the emergency shutdown button.

A team at Atomic Energy of Canada Ltd (AECL) have performed a detailed analysis of the Chernobyl power surge and the role of the graphite control rod followers. They have shown (Ref 2) that the immediate effect of the insertion of the fully withdrawn control rods would have depended crucially upon the conditions in the reactor at that time. The neutron flux spatial distribution at the moment of rod insertion is of particular importance and this depends on the irradiation history of the core and the positions of all the control rods. In most circumstances, the increase of reactivity at the bottom of the core would have been overwhelmed by the reduction in reactivity caused by entry of the control rods at the top of the core and the reactor would have shut down. Prior to the turbogenerator experiment however, xenon poisoning of the core and the withdrawal of almost all the control rods had caused an unusual double-humped vertical neutron flux profile to develop. The top and bottom halves of the reactor could therefore have acted to some degree independently. A transient local positive reactivity addition could then have contributed significantly to the severity of the power surge in the bottom half of the core and might even have initiated its rapid development. Without a more detailed knowledge of the exact conditions in the Chernobyl reactor at the time of the accident than is currently available, however, it is not possible to quantify the transient positive reactivity addition with any precision. Nonetheless it appears likely that the attempt to shut down the reactor did initiate such an event and that this contributed to the severity of the power surge. The Soviets have themselves estimated that, based on neutron flux distribution measurements made just before the accident, the magnitude of the positive scram effect was in the range 0 to 1.5β (Ref 9).

Some other factors might also have possibly contributed to the reactivity excursion by accelerating the development of voids in the coolant. Firstly, coolant flow in the fuel channels at the start of the turbogenerator test would have been in the bubbly flow regime. As the pumps ran down and the rate of steam generation increased, a flow pattern transition from bubbly flow to annular flow might have occurred, rapidly increasing the void fraction and hence reactivity of the core. Secondly, parallel channel instabilities might have occurred, causing rapid voiding of some channels although this could have been compensated by reduced voidage in other channels. Finally, interactions between the pumps that continued to operate and those that were running down might have caused the latter to stop prematurely, resulting in a more rapid flow reduction than anticipated. The pressure head developed by the pumps that continued to operate would oppose the continuation of flow through the pumps that were running down.

5.3 Failure of shutdown system

A crucial step in an accident is the shutdown of the fission reactions in the core. This action, known as either reactor trip or scram, is the first of a series of safeguards intended to prevent the accident progressing to the core degradation stage. It causes the rate of heat generation in the core to fall steeply. The Soviets have informed us that the operator at Chernobyl Unit 4 ordered a full emergency shutdown at 1:23:40 on April 26. All top-entry control rods were then motored into the core. Not all of them reached their lower stops however, so the operator uncoupled them so that they could fall "under their own weight". During this period, a number of shocks were felt, the pile cap (ie upper biological shield and charge face) was blown off and the reactor building was extensively damaged. Nuclear fission in the core then ceased.

It is clear that emergency shutdown was initiated and that nuclear fission in the core did eventually cease. It appears unlikely however, that shutdown was due entirely to insertion of the control rods. As the reactor went prompt critical, these could not have been inserted rapidly enough to terminate the reactivity excursion. In the view of experts at the IAEA Meeting, disintegration and dispersal of fuel in the accompanying explosions must have contributed to the shutdown. It therefore appears that major damage occurred before the reactor was shutdown. The premature arrest of the motion of the control rods into the core seems to support this. Furthermore, control rods were lifted when the pile cap blew off but fission still ceased, pointing to fuel dispersal and core damage as the reasons for shutdown.

COMMENT

It is clear that the Soviets were mistaken in considering multiple pressure tube rupture to be a 'beyond design basis accident' and failing to protect against it. This was a serious design error as the failure of several pressure tubes and the release of pressurized coolant could disable reactor protection systems and prevent shutdown of the 'chain reaction'.

Any design of reactor that permitted a reactor fault to disable a protection system in this way would not be acceptable in the UK as it would not conform to the NII's safety assessment principles 102, 108 and 123, which state:

"102. Safety-related structures and plant should be protected as appropriate from the radiation, thermal and dynamic effects of any specified fault involving the coolant;"

"108. The reactor and associated plant should be designed, constructed and operated so that the reactor can always be shutdown and held shutdown in a safe sub-critical state thereafter;"

"123. The protection system equipment should be designed, laid out and sited that, notwithstanding the effect of plant faults, adequate protective action will be available."

This conclusion is reinforced by the CEGB's design safety criteria, which require that (Ref 4, subsection 4.2):

"all potential hazards originating from within the power station site boundary shall be considered. Those hazards to be considered shall include fires, explosions, releases of gases, water, steam or noxious substances, failure of pressure parts, supports, or other structural components, disruptive failure of rotating machinery and dropped or impacting loads.

It will be necessary to ensure that hazards considered within this category do not give rise to conditions which might invalidate the protective systems and lead to an unsafe situation;"

"structures containing gases, liquids, or noxious substances shall be designed so as to minimise as far as reasonably practicable the possibility of sudden failure.

Safety related structures, systems, cabling and components shall be designed and located so as to minimise, consistent with other safety requirements, damage due to the release of gas, water, steam or any noxious substance.

Special care shall be taken to ensure that the release of any of the above substances will not prevent any necessary operator action to control the incident or to safely shut down and cool the reactor;"

"notwithstanding the preventive measures, the possibility that any container of the above substances may fail and release its contents shall be considered and where necessary the means to

control the effects of such releases shall be provided."

5.4 Core degradation and pressure circuit failure

The Soviets have told us that fuel degradation occurred while there was still water in the fuel channels, due to a heat transfer crisis at the surface of the fuel rods. The fuel then disintegrated and rapidly generated steam by mixing with the water present in the channels. The abrupt pressure pulse that this generated ruptured fuel channels and resulted in the 3m thick pile cap and floor above it being blown off and core materials being ejected into the atmosphere. Part of the emergency core cooling system capability was destroyed by the explosions.

COMMENT

Had the reactor been designed so that it always had a negative power coefficient and could not operate in any regime where it suffered instabilities on short timescales and had the reactor operators not been able to disable engineered safeguards and violate operating regulations prior to the experiment on the turbogenerator, the accident would never have happened. Once the accident had been initiated and the reactor had become prompt critical however, the delay between safeguard actuation and operation would have been too long to prevent core damage. Even if emergency core cooling had been available during the accident, it would not have significantly affected the course of the accident due to the heat transfer crisis at the surface of the fuel rods, which allowed fuel degradation to occur in the presence of water.

The events following accident initiation reveal a number of areas where the RBMK design is deficient from the UK viewpoint and reinforce the view that the design would have been unacceptable in the UK. First, it is noted that the fuel is in close

proximity to the boundary of the pressure circuit (pressure tubes), so that overheating of the fuel could have an adverse effect on the integrity of the pressure boundary and yet the design makes no special provision to safeguard against this.

For the UK however, the NII safety principle 95 requires that:

"95. Where overheated fuel could cause failure of the primary coolant circuit or where the fuel geometry could be so changed as to adversely affect the heat transport process it should be shown that adequate provisions have been made in the design to inhibit such a situation or that additional safeguards would be available to maintain the plant in a safe condition and to prevent any release in excess of the requirements of principles 13 to 17."

Furthermore, the release of pressurized coolant from some ruptured fuel channels blew off the pile cap causing the rupture of the remaining fuel channels and pressure tubes above the core, which destroyed part of the emergency core cooling capability.

This adverse interaction with the ECCS would not have occurred in a reactor designed to conform with the NII safety assessment principles 102 and 123 already mentioned in connection with the shutdown system, or NII safety principle 258, mentioned in connection with the propagation of fire to Unit 3.

It is known that pressure tubes failed in the accident at Chernobyl and their manner of failure is possibly of secondary importance for an appreciation of the consequences of their failure. Nonetheless, to demonstrate the more robust response of other reactor systems were a similar fuel disruption event to occur, it is necessary to understand the manner of failure at Chernobyl.

There are several related phenomena that could have caused disruption of the fuel, rupture of fuel channels containing water and the expulsion of the pile cap. These are summarised below. The feature common to all of these is that fuel fragments and its surface area for heat transfer increases, allowing very rapid steam generation.

A mechanism for fuel disruption was proposed at the IAEA Meeting. When high burn-up fuel is heated slowly, the pressure exerted by volatile fission products trapped on grain boundaries within it increases and this can cause the fuel to swell. If the fuel is heated very rapidly however, tensile stresses resulting from the pressure build-up cannot be relieved sufficiently quickly and the fuel can disintegrate into a fine powder, blown apart by the pressure of the volatile fission products. At Chernobyl, the fuel was heated rapidly. It was suggested that if the heating were rapid enough, the fuel might have disintegrated prior to melting and been expelled into surrounding water, resulting in a very rapid burst of steam generation.

More recently, fuel failure in the Chernobyl accident has been analysed by a USDOE team using the FPIN-2 fuel rod behaviour computer code (Ref 6). This predicted that fuel rods would have failed near the bottom of the core once the fuel had started to melt. Fuel melting causes a rapid increase of pressure in the fuel rods, because the fuel volume increases by 10% on melting, reducing the free volume available to accommodate fission gases. Experiments have shown that this fuel expansion causes rapid localised deformation of the hottest cladding, which quickly leads to cladding failure. The USDOE team's assessment was that at the time of cladding failure, 57% of the fuel in the vicinity of the failure would have begun to melt and that 10% of this would have actually been molten. Following cladding failure, molten fuel, unmelted fuel and fission gases would have been ejected into the fuel channel.

The first fuel channel ruptures were probably caused by steam explosions, resulting from either the ejection of fine particles of fuel debris into water remaining in the channels, or energetic molten fuel-coolant interactions. In the latter case, fuel and clad would melt, mix with water remaining in a fuel channel and then fragment into fine particles, rapidly transferring heat to the water and causing an abrupt rise of pressure. The general geometry of an RBMK fuel channel is reminiscent of some of the configurations that have been used in experiments to promote energetic molten fuel-coolant interactions.

Another mechanism that might possibly have ruptured fuel channels is the "steam spike". As for energetic molten fuel-coolant interactions, this requires fuel to melt and mix with water. Steam spikes differ from steam explosions however, in that heat transfer occurs over a more extended period (perhaps tens of seconds rather than milliseconds) and steam spikes are not accompanied by the shock waves and fine fuel fragmentation that are characteristic of steam explosions.

Possible thermal interactions of fine particles of fuel debris with water in fuel channels during the Chernobyl accident have been analysed by a USDOE team using the EPIC computer code (Ref 6). Particles with radii of 10, 100, 300 and 600μm were considered. These were assumed to be ejected from a fuel rod into a fuel channel over a timescale of several tens of milliseconds. The best estimate of channel pressurisation due to these thermal interactions was about 15MPa. Calculations were also performed with the ALICE-2 fluid-structure interaction code to assess the response of the fuel channels to these interactions. It was concluded that these thermal interactions would not have pressurised fuel channels sufficiently to cause ruptures within the core, although they might have caused failures close to channel top closure heads.

The most likely reason for the first ruptures of fuel channels is that energetic molten fuel-coolant interactions occurred in individual fuel channels. Indeed, this type of steam explosion has occurred in an experiment (Ref 7) under conditions similar to those in Chernobyl Unit 4. The USDOE analyses indicate that about 6% of the fuel in the lower of the two subassemblies in a fuel channel could have been molten at the time of fuel failure. This amounts to about 3 kg of molten UO_2. In the Scale-Urania-Water (SUW) series of molten fuel-coolant interaction experiments at AEE Winfrith (Ref 8) in which 24 kg masses of molten fuel simulant were released into pools of water, energetic interactions involving on average about 3 kg of melt were routinely observed to generate mechanical yields in the range 100-1000 kJ. The mechanical energy required to rupture a Chernobyl fuel channel is estimated to be of order 1 kJ, so an energetic molten fuel-coolant interaction involving a small fraction of the molten fuel released into a channel should have been capable of rupturing it. Furthermore, the fuel channel walls might have been weakened prior to the steam explosions by molten fuel impinging on them.

The participation of any molten zirconium alloy in steam explosions would have resulted in its fine fragmentation and rapid hydrogen generation. It has been suggested that a chemical explosion occurred three to four seconds after the pile cap was blown off and this would have required the rapid generation of combustible substances such as these. In steam explosion tests involving molten aluminium, subsequent chemical explosions involving the rapid oxidation of finely divided aluminium have often been observed.

Steam spikes appear to be a much less likely explanation of fuel channel rupture. As they occur over a more extended period than steam explosions, time is available for pressure relief within the system. Consequently a much larger water mass has to be vaporised to produce a similar pressure jump. Furthermore, heat transfer from the fuel to coolant becomes inefficient once the pressure has risen above the critical pressure of the coolant. The result of this is that the maximum pressure generated by a steam spike is generally assessed to be not much greater than the critical pressure of water. As the fuel channels at Chernobyl Unit 4 contained water during the reactivity excursion, the strength of the pressure tubes would have been close to normal and rough estimates indicate that they could have withstood the critical pressure of water. This suggests that a steam spike would not have ruptured fuel channels.

The Soviet analysis of the accident (Figures 1a-1c) indicated that the fuel overheated and disintegrated while still surrounded by water. Had the fuel channels dried out in some locations however, the pressure tubes might have weakened and failed due to thermal radiation from the fuel and fuel relocation leading to contact between the fuel and pressure tubes. The integrity of the pressure tubes would be in doubt once they reached a temperature of about 700C. Simple scoping calculations of transient heat transfer indicate that if heat were generated in the fuel at three times the normal operating rate, the fuel would have risen to the clad melting temperature (about 1850C) by the time the pressure tube wall had reached 700C. At higher powers, fuel melting would occur before pressure tube failure. The Soviet analysis indicated that the power in the Chernobyl reactor rose to 100 and perhaps 440 times the normal operating value and so both fuel and clad would probably have melted before thermal radiation raised the pressure-tubes to their bursting temperature of about 700°C. Nonetheless pressure tubes could still fail as a result of fuel relocation creating hot-spots on the pressure tube walls.

The core inerting system at Chernobyl Unit 4 was designed to relieve the pressure that would have resulted from a single pressure tube rupture. Following the rupture of several channels, the core space was pressurised and the pile cap was blown off, rupturing those fuel channels that were still intact and exposing the core to the atmosphere.

COMMENT

In this respect also, the RBMK design does not conform to the NII's safety assessment principles for number 152 states:

"152 A containment should be provided around the reactor and its primary coolant circuit, unless it can be shown that adequate protection has been achieved by some other means. The containment should adequately contain such radioactive matter as may be released into it as a result of any fault in the reactor plant."

The energy required to cause this damage could have been supplied by one of the three fuel-coolant interaction mechanisms described above, but it is more likely to have been supplied by the thermal energy already stored in the coolant in the fuel channels. Thermodynamic models of fuel-coolant interactions make no assumptions about the process of fuel fragmentation and can be used to estimate the mechanical yield of any of the three types of fuel-coolant interaction described above. If the mass of fuel participating in these interactions were equal to the mass of fuel expelled from the reactor vault (about 4% - similar to the mass in the smallest critical configuration in the core), mechanical work of order 1 GJ could have been done by fuel-coolant interactions. Even in the absence of a large contribution to mechanical yield from fuel-coolant interactions, however, the coolant in the fuel channels alone might do mechanical work of order 1 GJ in escaping from ruptured channels. These yields compare with rough estimates in the range 0.2 to 2.0 GJ, of the work done in blowing off the pile cap.

Following the steam explosions, the core cooling systems and all fuel channels were destroyed. It has been suggested that the expulsion of the pile cap then allowed hydrogen formed by the chemical reaction of steam with zirconium to mix with air in the reactor hall and explode, further damaging the reactor building (Ref 1). Prior to this, hydrogen would have been unable to burn or explode as there was no oxygen in the pressure circuit and oxygen was excluded from the reactor vault by inert gas blankets around the core. The accident then developed as a loss of coolant accident affecting the whole core.

5.5 Response to the accident

At Chernobyl, the containment was breached within seconds of accident initiation. Witnesses outside Unit 4 reported that burning lumps of material and sparks shot into the air above the reactor. It was likened to a fireworks display. Some of the burning material fell onto the roof of the turbine hall and started a fire that put Unit 3 at risk. Altogether, over 30 separate fires were started on the site. By 0130 hours the firemen on duty had been called out and were reinforced with fire fighting units from Pripyat and Chernobyl. Graphic accounts have been given of the extreme heroism of these firemen, many of whom have since perished as a result of their exposure to lethal doses of radiation. By 05.00 hours the fires on the reactor and turbine buildings had been extinguished. Amazingly the three other units at the station continued to operate. The No 3 unit, which was adjacent to the damaged unit, was not shut down until 05.00 hours. The other two units continued to operate until the early hours of the following morning, some 24 hours after the accident.

With the reactor core and large amounts of hot zirconium alloy exposed to the atmosphere and the cooling system wrecked, rapid exothermic oxidation of zirconium commenced. It is thought that this raised the temperature of the graphite and ignited it. The Soviet engineers then had to consider how to fight the fire and how to reduce core temperatures, deal with decay heat and limit fission product release. They initially tried to cool the core by use of the emergency and auxiliary feedwater pumps to provide water to the core. This was unsuccessful. A video recording of the top of the core with parts of it glowing red hot was shown by the Soviets at the IAEA Meeting. Given the graphite fire and the continuing release of fission products, the decision was taken to cover the core with various materials: 40 Te of boron carbide was dropped first to ensure shutdown; then 800 Te of dolomite (limestone) was dropped. This releases carbon dioxide as it heats up and decomposes, which would starve the core of oxygen. 2400 Te of lead was then dropped, in the hope that this would melt, run through the core debris and carry heat away. Finally, clay and sand were dropped to filter and retain fission products escaping from the core. Altogether, about 5000 Te of material was dropped onto the core. The reactor core was thus covered by a loose mass that filtered the fine aerosol fission products.

From early May the position at the damaged reactor improved. Monitoring devices to measure temperatures and air speeds were lowered into the debris. The fuel temperature varied in a non-monotonic manner (see Figure 2) following the accident. At the moment of the explosion, the fuel temperature had reached 3000 C (at least locally). Immediately afterwards, the effective temperature of fuel remaining in the core was assessed to be 1300-1500 C and this fell over several tens of minutes as heat diffused into the graphite. Subsequently, the fuel temperature increased again over a period of several days due to fission product decay and the thermal insulation provided by the materials dropped onto the core. By 4-5 May, the fuel temperature had reached about 1900 C , as would have been expected from these two effects. It then began to decrease, however. The Soviets say this was the result of improved core cooling by circulating air and nitrogen injection. However it could be that the nitrogen extinguished fires in the graphite and zirconium and that this helped reduce the temperature. None-theless, the decay heating rate at this time would still have been substantial (about 5MW). Another possibility is that the fuel remelted and poured out of the core on to the concrete below.

Long term control measures have been instigated by the Soviets. The pressure suppression pools beneath the reactor have been drained and a flat heat exchanger embedded in concrete has been placed below the building foundations. Unit 4 has been entombed in protective concrete walls 1m thick. This includes a perimeter wall enclosing the turbine and reactor blocks as well as internal dividing walls between Units 3 and 4 and a protective cover over the turbine and reactor blocks. An internal recirculating ventilation/cooling system has been installed and the entombed reactor will be maintained at reduced pressure (with respect to atmospheric pressure) and the exhausted air discharged through filters and a stack.

The state of the core following the explosions is uncertain. It is known that the pressure tubes leading from the tops of the fuel channels separated from them when the pile cap was expelled, but the state of the pressure tubes leading to the bottom of the core is unknown. Fuel would not have been lifted out of the core by the rising pile cap, because when the fuel channels ruptured, the pile cap would have accelerated so rapidly that the tubes supporting the fuel assemblies would also have snapped. Being no longer suspended from the top of the core, the fuel rods would have tended to fall towards the bottom of the core. Fuel would not have been able to migrate into the lower part of the reactor vault unless molten or fragmented but it is suspected that much of it is now there.

The core temperature would then have gradually risen, due to decay heating and oxidation of core components. The materials dropped onto the core would have thermally insulated the core and inhibited ingress of air. The core temperature of about 1900 C on 4-5 May would have been expected from decay heating of a well insulated core. Dropping lead onto the core would not have had a great effect on this heat-up, as the melting point of lead is relatively low (328 C) and its thermal capacity would have been about an order of magnitude less than that of the graphite. It could, however, have plugged the inlet piping if this were still intact and prevented cooling by circulating air or gases along the pipes.

The Soviet explanation of the arrest and subsequent fall of core temperature on 4-5 May is somewhat puzzling. They have said that the temperature began to fall due to the formation of a stable convective air flow through the core into the free atmosphere and pumping nitrogen into the space beneath the reactor. A scoping calculation of free convection through a fuel channel, open at both ends, indicates that the fuel would have had to heat the air to about 1000 C to create a sufficient draught to remove the decay heat. At this temperature, the zirconium alloy clad and graphite would have burned and exacerbated the heat rejection problem. Furthermore, the fuel channels were not open to the atmosphere at both ends: the core was covered by several metres of dolomite, sand and clay, and lead that had melted and run through the core might have plugged the inlets. Thus, still higher fuel temperatures would have been required. To remove the decay heat by gaseous nitrogen injection would have required an injection rate of about 0.5 Te per minute.

A more likely explanation would have been that the zirconium alloy pressure tubes and/or stainless steel inlet pipework melted on 4-5 May and fuel debris fell from the graphite onto the concrete floor of the reactor vault. Concrete melts at about 1400 C, so core-concrete interactions could have brought about the fall in core temperature. Lead boiling could also have played a role (boiling point of lead = 1740 C). The Soviets have said however, that the core debris did not interact with the concrete and that the maximum temperatures in the reactor vault are now only several hundred degrees Celsius.

5.6 Summary of comments

The accident at Chernobyl has revealed numerous deficiencies in the design of RBMK reactors when judged by the NII's safety assessment principles. These principles are not theoretical ideals that would be unobtainable in practice, but rather "are based on experience obtained so far on the operation of commercial plant in the United Kingdom and represent a level of protection against radiological consequences of normal operation and fault conditions that should in most circumstances prove

to be reasonably practicable" (Ref 3). The design deficiencies would have made RBMK reactors unacceptable in the UK.

There appear to be design deficiencies in six main areas:

I positive power coefficient;

II inadequate shutdown system;

III scope for operator interference with safeguards;

IV instrumentation and alarms;

V propagation of damage on-site;

VI vulnerability of safeguards to fault conditions.

These are outlined below.

I Positive power coefficient

The RBMK reactor normally has a positive void coefficient of reactivity that is "held in check" by the negative Doppler (temperature) coefficient. When operating below about 20% full power however, the void effect on reactivity can dominate the Doppler effect and the power coefficient of the reactor can become positive. Furthermore, the reactor becomes subject to thermal-hydraulic instabilities at low power (Ref 5), which cause coupled power fluctuations. A core design with these characteristics would not conform to the NII safety assessment principles numbers 68 and 93, which require a design to be such that these characteristics are avoided or minimised. The NII requires reactor designs to conform to its safety assessment principles "as far as is reasonably practicable" and this was clearly not the case for RBMK reactors as the Soviets now intend to rectify this deficiency by increasing the fuel enrichment and number of absorber rods in them (Ref 5). The RBMK core design would have been unacceptable in the UK.

II Inadequate shutdown system

The RBMK reactor relies upon a single system to achieve shutdown following a fault, so a failure of that system could prevent the reactor being shutdown. This aspect of the design does not conform to one of the NII's general engineering principles for nuclear plant design (No 31) and does not conform to several specific engineering principles (112, 115, 121, 122), which require diversity of protection systems where there is doubt about the effectiveness of a non-diverse system. There clearly was doubt about the effectiveness of the shutdown system: it was known to be inadequate when the "reactivity reserve" fell below the equivalent of 15 fully inserted rods and for this reason, operation with a reactivity reserve of less than 15 rods was forbidden. This second design deficiency has now been recognised by the Soviets, who are evaluating additional diverse and fast-acting shutdown systems. The NII principles further require the designer to recognise that maloperation may occur and to design so as to ensure that any limits placed on components of a protection system are not infringed. The RBMK design was also deficient in this respect. It would not have been acceptable in the UK. The possibility of a transient increase in reactivity caused by the activation of the shutdown system would violate NII principle 143. The Soviets have now modified the control rod/follower system on their RBMK reactors to prevent the lower ends of the followers rising above the bottom of the core, thus preventing any possibility of "positive scram".

III Scope for operator interference with safeguards

NII safety assessment principle 107 states:

"107. Adequate protective systems should be provided and, whenever fuel is in the reactor, they should be maintained at a level of readiness adequate to ensure nuclear safety."

During the preparations for the turbogenerator experiment, operators disabled some systems and blocked trip signals to others. The NII foresaw in its safety assessment principles that such interference might be attempted and so formulated several principles (38, 127, 131) that require a certain minimum amount of operational protection equipment to be specified. The design must then be such that access to this equipment is "physically controlled" to ensure availability of this minimum. The RBMK design clearly did not conform to these safety principles.

IV Instrumentation and alarms

At the start of the turbogenerator test, steam voidage in the core was tiny and conditions were such that a rapid and large increase of voidage could occur with a consequent adverse effect upon reactivity. The operator did not, however, appreciate the state of the plant. The Soviets have subsequently recognised this design deficiency of RBMK reactors and now intend to install additional instrumentation to measure coolant subcooling at the main coolant pump inlets. The importance of providing plant operators with clear and comprehensive information on plant state has been recognised in the UK for many years however, and embodied in the NII safety assessment principles (66, 132). These principles also require that alarms be given when set limits are reached (principles 132 and 129). The RBMK design was deficient in this area too. The information that the reactor had a reserve reactivity margin below the lowest permissible limit at the start of the turbogenerator test was probably buried amongst other data on a computer print-out that itself was only generated by a request from the operator. Information of such importance should not await a demand from the operator. The level of instrumentation and alarms on RBMK reactors would clearly have been unacceptable in the UK.

V Propagation of damage on-site

The danger of fire travelling along the roof of the turbine hall to threaten the adjoining Unit 3 forced the firefighters first of all to concentrate on extinguishing the fires on the turbine hall roof. It was during this action that many of them received fatal doses of radiation. Had the roof been made of non-flammable materials, this danger would have been reduced. In the UK, the effect of plant layout and building materials on the potential for accident propagation is considered as thoroughly as other aspects of nuclear plant safety and this is reflected in related NII safety assessment principles (in particular, 258). The CEGB goes further still by requiring (Ref 4) that "non-

combustible and heat-resistant materials shall be used wherever practicable throughout," and "where it is impracticable to use non-combustible materials, then the use of installations and materials specifically designed to reduce the rate of propagation of fire shall be used." The RBMK design did not conform with these principles.

VI Vulnerability of safeguards to fault conditions

Finally, the pressurisation of the space beneath the pile cap by coolant released from the fuel channels caused the pile cap to be blown off, lifting control rods out of the core and destroying part of the emergency core cooling capability. The Soviets mistakenly considered multiple pressure tube failure to be a 'beyond design basis accident' and failed to protect the reactor against it. Any reactor design that permitted a fault to disable protection systems in this way should be unacceptable in the UK as it would not conform to several of the NII's safety assessment principles (102, 108, 123), which require that protection systems be safeguarded against the effects of plant faults in general and faults involving the coolant in particular.

The design deficiencies in any one of these six areas would have rendered the RBMK reactor unacceptable in the UK.

5.7 Conclusions

It is now clear that the Chernobyl accident was in essence a fuel-coolant interaction triggered by a "prompt critical excursion". The prompt critical excursion caused a rapid power surge. It was due to a turbogenerator experiment causing a reduction of coolant flow and increased boiling, coupled with a fundamental instability at low reactor power. This caused a spontaneous increase in the power to cause still further increases of power due to a positive feedback phenomenon. Furthermore, it appears likely that a serious flaw in the design of the shutdown system significantly increased the severity of the power surge by causing a transient positive reactivity addition in the lower part of the core when the emergency shutdown button was pressed.

Some press reports have described this rapid power surge as a "nuclear explosion" but that is incorrect. To achieve a nuclear explosion from an atomic bomb requires bringing together the nuclei capable of undergoing fission (uranium 235 or plutonium 239) very rapidly and keeping them together for long enough for a significant fraction to undergo fission. Weapons designers do this by driving the fissile components of the bomb together with explosive force, producing rapid increases in reactivity far beyond the 'prompt critical' state. This can only be achieved by using uranium or plutonium with a very high proportion of the fissile isotopes. In these circumstances the fissions are due to fast neutrons, the time between successive fissions is very short and massive amounts of energy are released before the material has time to blow itself apart and thus terminate the fission chain reaction.

In a thermal reactor the situation is very different. The fuel is low enriched uranium in which the fissile isotopes (uranium 235 and some plutonium 239) only represent about 2 to 3% of the total, the remainder being uranium 238 which absorbs neutrons. The fuel is also intimately associated with a moderator that acts to slow down the neutrons produced by fission. In a power surge, even one in which a 'prompt critical' state is reached, the presence of the uranium 238 would reduce the increase in reactivity by absorbing more neutrons as they are slowed down - the Doppler effect. More importantly, the majority of fissions would be caused by slow neutrons. The time between successive fissions would therefore be longer than is the case with fast neutrons and the energy produced would disrupt the fuel and hence terminate the fission chain reaction long before the reactivity reached the very high levels achieved in an atomic bomb. The proportion of the uranium or plutonium undergoing fission and hence the energy released would be far lower, volume for volume about 100,000,000 times less than would be achieved in a bomb. This is an intrinsic characteristic of all reactors including the fast reactor in which the neutrons have speeds intermediate between those associated with thermal reactors and atomic bombs.

It is theoretically possible to have a rapid power surge in a gas-cooled reactor or a PWR if a situation were postulated in which control rods were rapidly ejected from the reactor. In a gas-cooled reactor or PWR, the inherent characteristics of the reactor would compensate for the power surge and terminate the event. The special feature of the Chernobyl reactor was the potential for positive feedback because of the inherent instability of the reactor concept. The power surge at Chernobyl was therefore of exceptional severity because of the inherent characteristics of the reactor but it is incorrect to describe it as a nuclear explosion.

In the light of the Soviet information, it is seen that the accident happened because of:

i faults in the concept of the reactor (inherent safety not built-in);

ii faults in the engineering implementation of that concept (insufficient safeguard systems);

iii failure to understand the man/machine interface ("a colossal psychological mistake" in the words of Mr Legasov);

iv poor operator training.

These faults placed an intolerable burden of responsibility for the safety of the nuclear reactor upon the operators. They are symptomatic of the differences between the Soviet and UK approaches to nuclear safety, which have been highlighted throughout this Section by comments drawing attention to the many aspects of RBMK design that do not conform to specific safety assessment principles enunciated by HM Nuclear Installations Inspectorate. It is instructive to see how they would have been avoided by the application of UK nuclear safety principles.

i First, and most important, in UK reactors protection against reactivity faults is achieved by selecting design concepts with inherent characteristics which provide built-in protection. For example, UK gas cooled reactors do not have a void coefficient either positive or negative and the void coefficient in a PWR is either negative or so slightly positive that the power of a reactor as a whole cannot run away. However, the Soviet RBMK reactor has such a large positive void coefficient at low power that it can dominate the behaviour of the reactor and the reactor can be intrinsically unstable; that is, if the power increases spontaneously, then it increases still more and escalates to higher and higher values. This positive feed-back phenomenon is not an inherently safe characteristic.

ii Secondly, in UK reactors, the natural defences provided by the intelligent choice of reactor concept are supplemented by engineering features to prevent, limit, terminate and mitigate faults. Thus, for example, it is physically impossible to withdraw the control rods rapidly and if the operator seriously mishandles them, then the reactor automatically fails safe, ie shuts down. In other words, the engineering implementation of a UK design ensures that the reactor remains safe even if the operator tries to do the wrong thing. The Soviet reactor was the exact reverse of that. The Soviet designers knew that their design was intrinsically unstable at low power, they knew that was potentially unsafe but they did not take any engineering steps to avoid that unsafe condition. They simply instructed the operators not to operate the reactor below 20% power and they relied upon the operators to follow that instruction faithfully.

Furthermore, in any reactor and especially one which was intrinsically capable of rapid positive feedback and power excursions, UK safety rules would insist on fast-acting control rods. The Soviets also neglected to provide these.

iii Thirdly, for UK reactors, the man/machine interface is assessed very carefully. The information to be given to the operator in the control room is considered, the reactor is designed so that he does not have to make important decisions in a hurry and fail-safe devices are fitted so that if the operator does do the wrong thing, the reactor fails safe. The Soviets admitted that they had not previously realised the importance of these points.

iv Fourthly, in the UK, operator training is regarded as extremely important. The CEGB and SSEB insist that their operators be highly qualified and have regular refresher courses including training on simulators. The Soviets admit that their training was inadequate.

v Finally, the commercial UK nuclear system is overseen by an independent nuclear inspectorate, which can at any time without hindrance or challenge close down any licensed reactor. The Soviets do not appear to have had an effective independent inspection capability.

In conclusion, the Chernobyl accident was so unique to the Soviet RBMK reactor design that there are very few lessons for the United Kingdom to learn from it. Its main effect has been to reinforce and reiterate the importance and validity of UK safety standards. A large scale reactor accident of the type that occurred at Chernobyl could not happen in the United Kingdom.

References

1 "The accident at the Chernobyl' nuclear power plant and its consequences", information compiled for the IAEA Experts Meeting, 25-29 August 1986, Vienna, by the USSR State Committee on the Utilization of Atomic Energy.

2 "The Chernobyl accident: multidimensional simulations to identify the role of design and operational features of the RBMK-1000", P S W Chan, A R Dastur, S D Grant, J M Hopwood and B Chexal, Proc. Int. Top. Conf. on PSA and Risk Management, Zurich, August 30 - September 4 1987, Vol III, p1116.

3 "Safety assessment principles for nuclear reactors", HM Nuclear Installations Inspectorate, HMSO, 1979.

4 "Design safety criteria for CEGB nuclear power stations", CEGB Health & Safety Department, HS/R167/81 (Revised), 1982.

5 "INSAG summary report on the post-accident review meeting on the Chernobyl accident", Vienna 30 August - 5 September 1986.

6 "Report of the US Department of Energy's team analyses of the Chernobyl-4 atomic energy station accident sequence", US Department of Energy, DOE/NE-0076, (1986).

7 "Molten fuel-coolant interaction occurring during a severe reactivity initiated accident experiment", M S El-Genk, NUREG/CR-1900, EGG-2080, 1981.

8 "An experimental study of scaling in core melt/water interactions", M J Bird, 22nd National Heat Transfer Conference, Niagara Falls, Paper Ref 84HT17, 1984.

9 "The Chernobyl nuclear power plant accident: 1 year after", V G Asmolov et al, IAEA Conference on Nuclear Power Plant Performance and Safety, Vienna, 28 September - 2 October 1987. IAEA-CN-48/63.

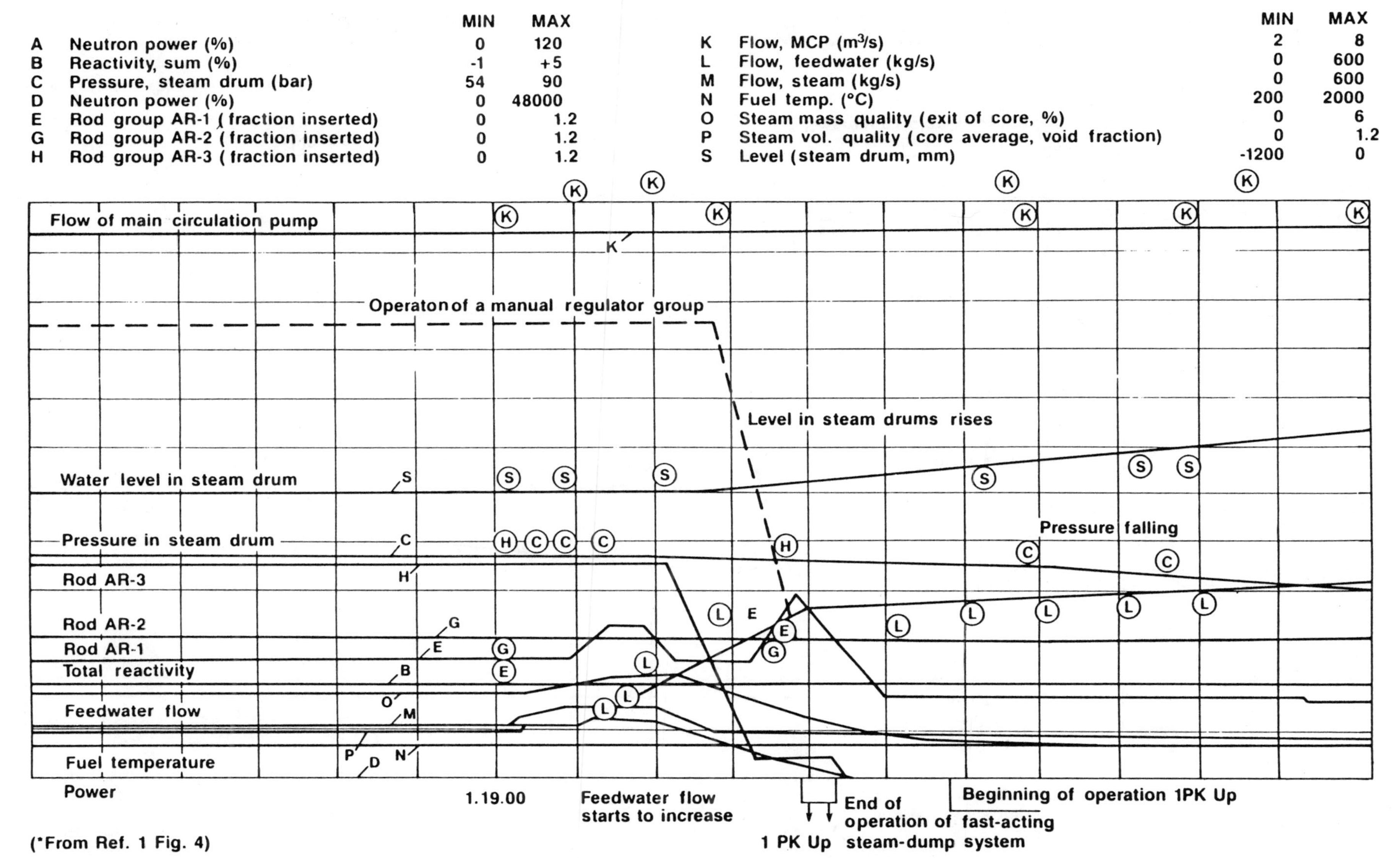

(*From Ref. 1 Fig. 4)

FIG. 1a USSR ANALYSIS OF THE INITIATION OF THE ACCIDENT AT CHERNOBYL - 4*

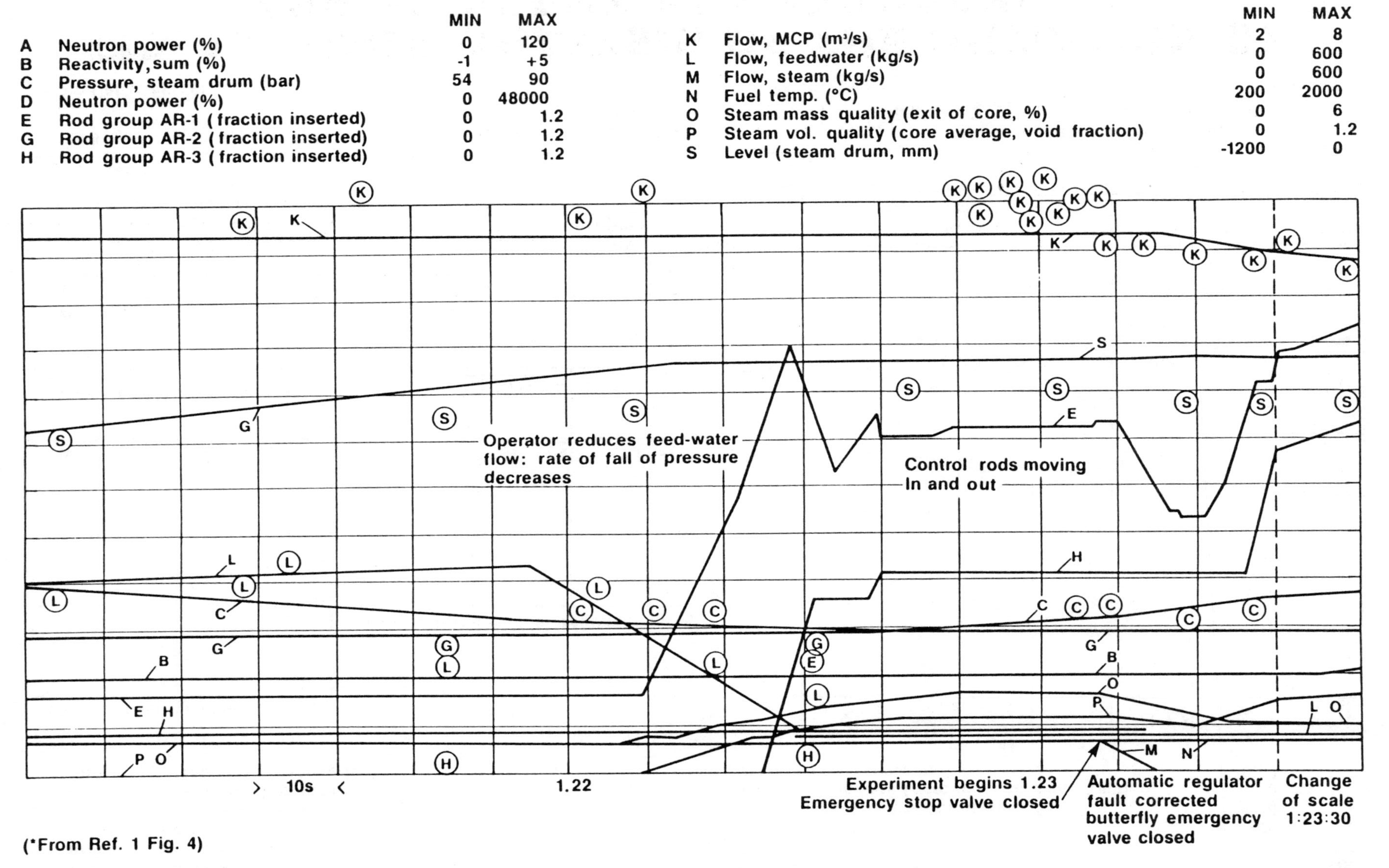

(*From Ref. 1 Fig. 4)

FIG. 1b USSR ANALYSIS OF THE INITIATION OF THE ACCIDENT AT CHERNOBYL - 4*

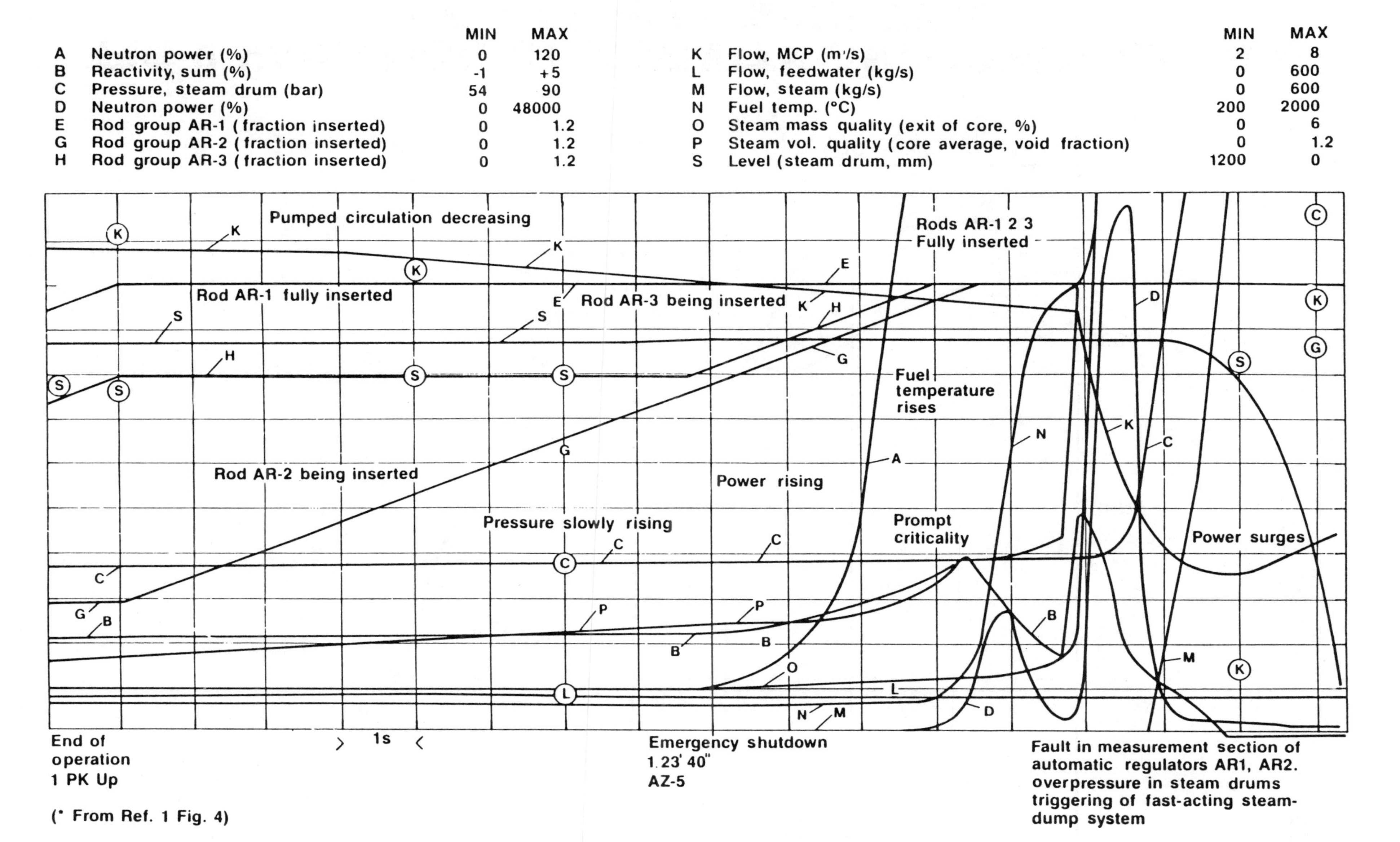

FIG. 1c USSR ANALYSIS OF THE INITIATION OF THE ACCIDENT AT CHERNOBYL - 4*

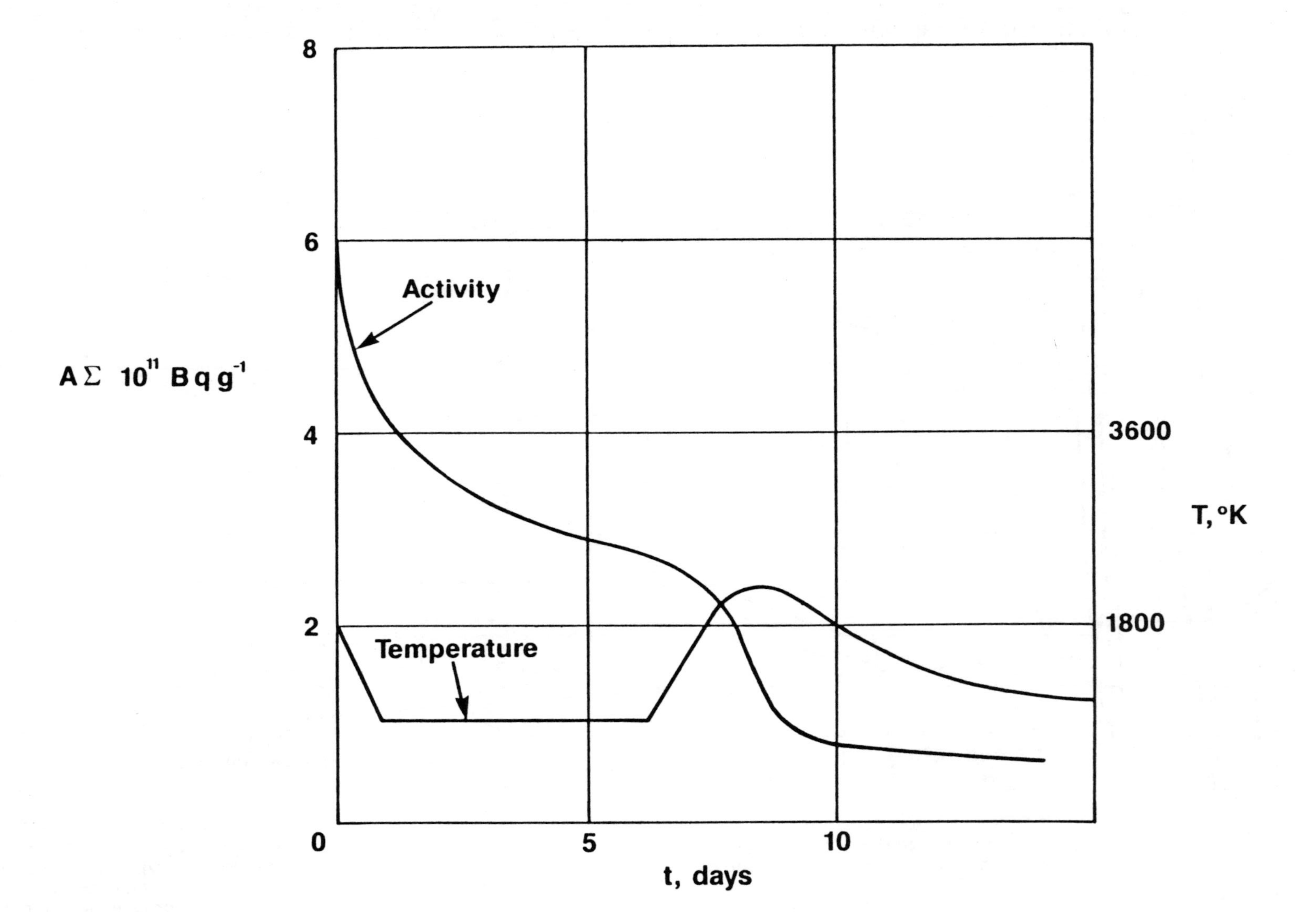

FIG 2 VARIATION OF ACTIVITY AND TEMPERATURE OF THE FUEL WITH TIME

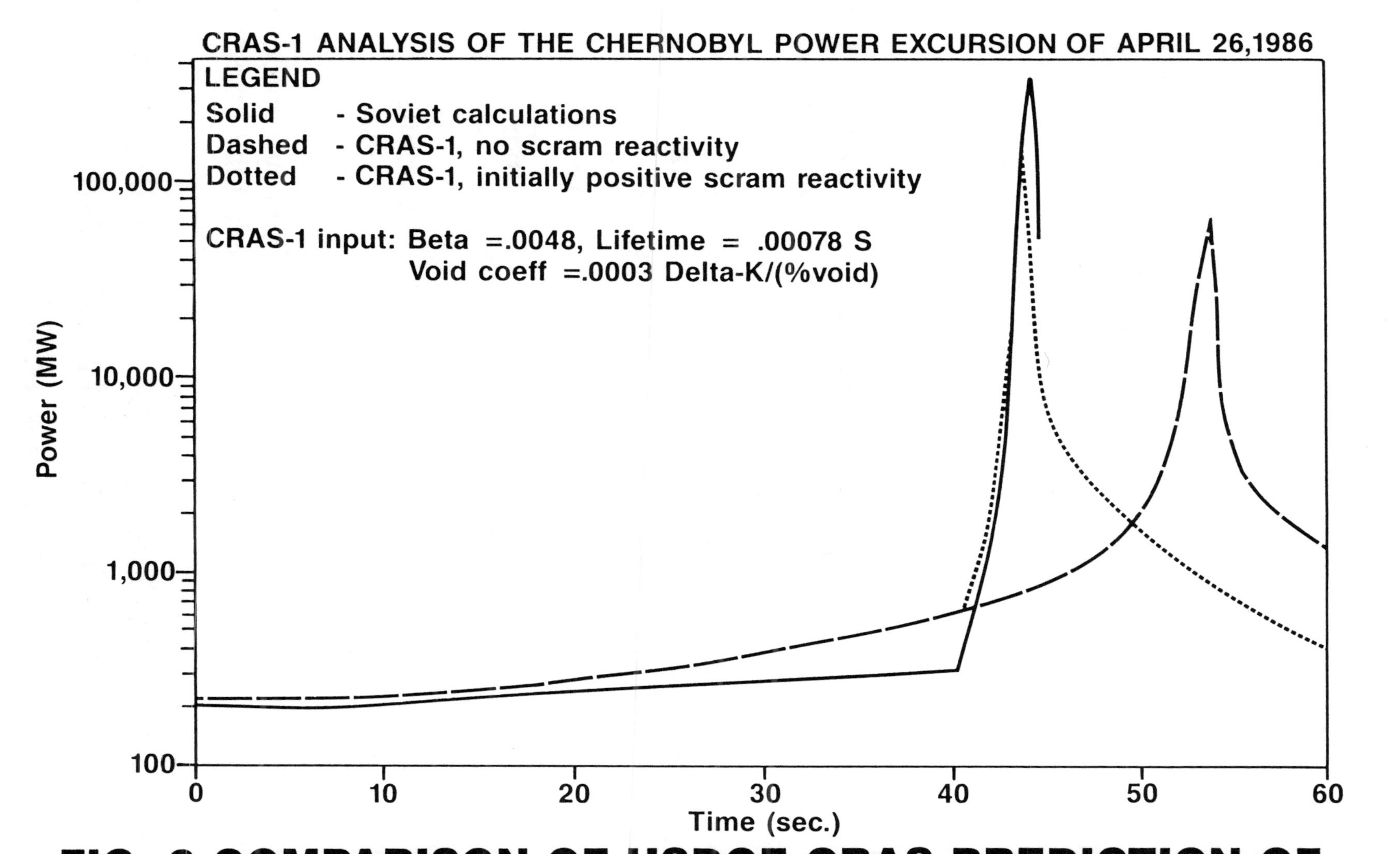

FIG. 3 COMPARISON OF USDOE CRAS PREDICTION OF POWER SURGE WITH SOVIET PREDICTIONS

(From USDOE report, DOE / NE-0076 Figure 12.a)

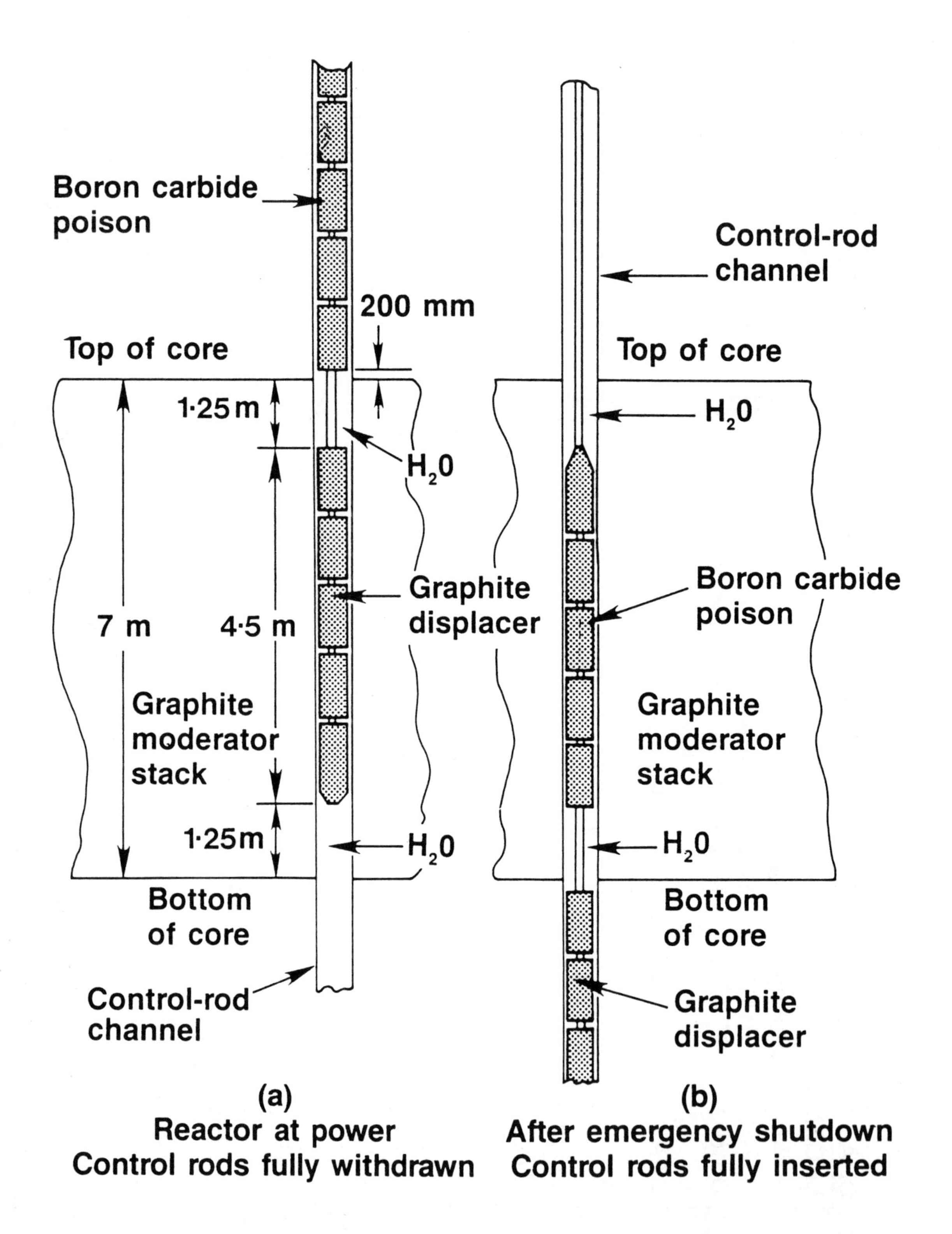

FIG. 4 POSITIONS OF CONTROL RODS
(a) FULLY WITHDRAWN
(b) FULLY INSERTED

SECTION 6: SOURCE TERMS

The source term describes the details of the radioactivity release associated with an accident which are required in order to assess the environmental consequences. The source term has several components:

a the amounts of the various radionuclides released to the environment, expressed as fractions of the initial core inventories (these can be converted into activities released given the activity inventories);

b the height and the energy of the release (these affect the height to which the plume rises and hence the distance the material is transported);

c time scales, such as the time and duration of the release and the warning time that a release is imminent.

Source terms are routinely calculated in safety studies for hypothetical reactor accidents. A selection of those calculated for the Sizewell B PWR Safety Study by the Westinghouse Corporation are exemplified in Table 1, which gives release fractions for the various classes of fission product. Four broad categories of accident are considered on this table:

Release Category A

Accidents of the severest type in which the containment fails or is bypassed by a leaking pipe at the moment when the core becomes completely molten. A substantial amount of highly active, volatile radionuclides will then be released to the environment.

Release Category B

Degraded Core accidents in which the containment fails some hours after the core melts: the release of radionuclides would be substantial although radioactive decay would have reduced their activity whilst dissolution, plate-out and aerosol sedimentation would have resulted in much retention.

Release Category C

All degraded core accidents in which the containment leaks or is penetrated below ground level but does not fail above ground.

Release Category D

In this category we might place all degraded core accidents in which the containment does not fail, but may nevertheless leak at the design-rate. The release of activity is then only small.

The primary source of information for the account of the Chernobyl source term given here is the Soviet report on the accident released at the IAEA meeting in Vienna in August, 1986 (1). Annex 4 of this report provides details of the Soviet source term data. The only subsequent Soviet source term information was provided at the IAEA Conference on Nuclear Power Plant Performance and Safety, Vienna, 28 September - 2 October 1987 (2). The Soviet sources are confined to considering only that part of the released radioactivity which settled within the boundaries of the USSR. This updated description of the Chernobyl source term takes account of the new Soviet information, of accounts and interpretation produced in other western countries, and of new data on the proportion of the released radioactivity transported outside the USSR. The description is divided into four topics: the history of the release, the inventory in the core at the time of the accident, the radionuclide composition of the activity sampled in various locations in terms of relative release fractions, and finally the absolute activity releases in the accident. Conclusions are drawn in a final section.

In severe accident studies it is usual to express the source terms by means of the fractions of the original core inventories of elements which are released (3). The data available on the Chernobyl accident are expressed in terms of the activities of individual nuclides which were either emitted from the plant on a given date, or were still present in the environment on a given date following earlier release and subsequent decay. In order to relate these to release fractions, it is important to use the notional core inventory of the nuclide in question corrected for decay between shutdown and the appropriate date as the divisor. For example, Table 8 shows that 0.15 EBq (1 EBq = 1 exaberquerel = 10^{18} Bq = 27 M Ci) of Te 132 was released on day 0 (26 April), representing a release fraction of 3.7% at that date. By day 10 (6 May), the cumulative activity of Te 132 in the environment from the initial and all subsequent release amounted to only 0.048 EBq, since this isotope has a fairly short half-life. The cumulative release fraction is then obtained using the notional core inventory of Te 132 on day 10 allowing for decay since shutdown, yielding a value of 9.8%. This applies to all the cumulative release fractions in the penultimate column of Table 8. Similarly, the total activity releases in Figure 1 are not actual day-by-day values, but are corrected for decay between the release date and day 10.

6.1 The History of the Release

Figure 1 shows the day-by-day total activity releases in the accident quoted in the initial Soviet account (1). The noble gases are excluded, and the uncertainty is stated as ± 50%. As explained above these activities are corrected to day 10 (6 May). This means that if all the core material in the environment were swept up on day ten and divided into piles according to what day the material left the reactor, the activities of the piles on that day would be the values on the histogram of Figure 1. The initial Soviet account identifies four stages of release:

Stage 1. (Day 0). This was the initiating in-core explosion at 01:24 on the 26 April. Significant quantities of fuel were ejected in the transient which blew off the pile cap, and these were accompanied by enhanced releases of the volatile fission products iodine, caesium and tellurium.

Stage 2. (Days 0-6). Following the initial explosion, the core top was left fully exposed to the atmosphere, and it appears that an intense graphite fire rapidly developed. High levels of activity release were associated with this during the first day. During the 27 April, dumping of materials onto the core debris from the air began to extinguish the fire, and by the 10 May some 5000 te had been dropped. In addition to boron compounds and lead, the materials included dolomite, clay and sand specifically intended to trap and filter out the active species from the core. This approach was reasonably effective, because during the 27 and 28 April the activity release rate was substantially reduced, and remained at a reduced level until the 2 May. However, release during this period was still very significant, and it is unclear whether the in-core fire was fully extinguished.

Stage 3. (Days 7-9). A marked increase in activity release occurred on the 3 May, which continued through the 4 and 5 May. This is attributed to the decay heat pushing the temperature of the core debris (fuel and moderator) to elevated levels, ultimately in excess of 2000°C, following the effective insulation of the pile up by dumped material. During the early part of this stage, strong enhancement of the volatile fission product contribution to activity release, notably of iodine, was observed. However, by the 5 May when the temperature peaked, the released activity assumed a composition close to that of the fuel itself.

Stage 4. (Day 10). On the 6 May, the release rate rapidly fell to effectively zero. This is attributed to a combination of factors which somehow brought about a rapid cooling of the fuel debris. A key feature seems to have been the injection of nitrogen at a high rate underneath the core, which simultaneously cooled the debris and stifled any residual graphite fires. Additionally, there is reference to special measures taken to promote the formation of more refractory fission product compounds by the introduction of further materials into the filter bed, but this is not enlarged upon.

Some additional detail on the early history of the release from the 26 to the 30 April is given in the second Soviet account (2). According to this, the early decline in the total daily activity discharge was considerably less abrupt than that stated previously. The revised estimate is shown in Fig 2 in terms of relative daily releases. No new information on the absolute magnitude of the total releases is given. Daily histories of the local fall-out rates of total gamma activity, and of Cs134, Cs137 and Ce144, within a zone extending to 80km around the reactor are also given. Some of these results are shown in relative terms in Fig 2. The pattern of the local fall-out is quite different from that of the total activity release, showing a distinct maximum on the 28 April, and falling steeply thereafter. The updated Soviet account states that local fall-out from the damaged reactor ceased within 4-5 days.

Some aspects of this new Soviet information appear to be difficult to reconcile with their original version of the history of the releases. For example, the data in reference 1 show that some 30% of the total releases of Cs134 and 137 occurred on the 26 April (see Table 8). This is not consistent with the deposition pattern shown in Fig 2. However, the two sets of information may not be contradictory if it is assumed that much of the caesium released during the first two days was in a form which did not settle out locally. This picture is in accord with the data on long-distance transport of caesium discussed below.

6.2 Activity Inventories in the Core

The mechanisms of release determine the fractions of the different elements released from the core. To calculate the absolute activity released the fractions have to be multiplied by the core inventories of the different nuclides. Conversely in an actual accident when activities are measured, these have to be divided by the inventories to obtain the information relevant to the mechanisms of release. Using the average burn-up of 10,300 M Wd/te, 2% fuel enrichment, and power history for the Chernobyl-4 reactor prior to the accident supplied to the 1986 Vienna conference (1), the UKAEA computer code FISPIN has been used to calculate the inventories at shutdown and through the 10 days afterwards. Neutron characteristics specific to an RBMK reactor derived from a WIMSD model were employed (3). The results for days 0 and 10 are shown on Tables 2 and 3 respectively. Nuclides are grouped according to the groupings of the Reactor Safety Study (4), and only those of importance for consequence assessment are included. At shutdown, ^{239}Np accounts for 20% of the activity.

The inventories quoted in Tables 2 and 3 are not definitive, since no allowance has been made for the variation in fuel burn-up in different regions of the core. The Soviet information states that some fuel with a burn-up extending to 15000 MWd/te was present. Comparison with other western calculations of shut-down inventory which include a burn-up distribution shows agreement within 10% with the results for individual fission products given in Table 2. However, considerably greater uncertainties are associated with the inventories of the actinides above mass number 240 due to the sensitivity to the assumed neutron energy spectrum (5).

The Soviet information does not contain explicit data on the Chernobyl core inventory at shut-down. However, it is possible to reconstruct this from the activity releases and release fractions for a set of key radionuclides quoted in Table 4.14 of reference 1. An inventory reconstructed in this way is compared with that calculated by FISPIN in Table 4. The general agreement is reasonably good, but there are some exceptions. Compared with FISPIN, the Soviet inventory significantly underpredicts the amounts of Te132 and Np239, and substantially overpredicts that of Ru106. Some direct evidence on the accuracy of the respective calculations is available from atmospheric sampling data on ratios of isotopes of the same element. The release and transport of such isotope pairs should preserve the ratios in the core, unless one of the isotopes is a daughter of an isotope of a different element which might be transported differently. The FISPIN and Soviet calculated inventory ratios for three elemental isotope ratios are compared in Table 5 with a variety of sampling results. The sampling data show considerable variability, and for both caesium and ruthenium isotopes, the FISPIN and Soviet values bracket the measured ratios. Both sets of calculations predict lower than observed Ce 144/141 ratios. On the whole, the sampling data variability is too large for conclusions on the relative merits of these calculations to be drawn.

6.3 Relative Release Fractions of Different Elements

The data on the composition of the various samples studied in the Soviet Union and elsewhere are here expressed in terms of "relative release fractions", defined as

$$\text{RRF} = (\text{release fraction of nuclide})/(\text{release fraction of } ^{137}\text{Cs})$$

$$= \left[\frac{\text{activity of nuclide in sample}}{\text{activity of nuclide in the core}}\right] / \left[\frac{\text{activity of } ^{137}\text{Cs in sample}}{\text{activity of } ^{137}\text{Cs in the core}}\right]$$

Each of the four activities which go to make up the RRF must be corrected to the same time. The result is then independent of time because the activity in any sample of any nuclide decays at the same rate as the total activity inventory of that nuclide. For this study, the FISPIN inventories are used to calculate the relative release fractions.

Relative release fractions for a variety of different samples are listed on Table 6. The first column is for the average of Soviet soil samples between 1.5 and 30 km of the reactor (Ref 1, Table 4.2). I, Cs and Te have effectively equal releases, and the non-volatiles (except perhaps actinides) have releases 10 - 20% of the volatiles. Put another way the samples look to be of fuel composition except for an enhancement of volatiles by a factor of around ten. Columns three to six refer to the aerosol airborne above the reactor on various days (Ref 1, Table 4.10). The day 1 and day 3 pattern is similar to that of the average soil samples, except that the volatiles enrichment is lower - more like a factor of 3 rather than 10. Column five is an average over the four days, 6 - 9, the time when the Soviet account speaks of excess volatiles being released. This excess does not show up in the average, which is broadly similar to the day 3 pattern. Only the day 7 sample (column six) shows a significant enhancement of volatiles, with the non-volatile RRFs down to a few percent, except for Mo and Ru, which stay up at around 30%.

Some independent evidence shows a substantial proportion of non-volatile material in the release. Columns seven and eight of Table 6 refer to measurements performed in the UK on specimens retrieved from regions relatively close to the reactor. Column seven gives the activity composition on the shoe of a British student evacuated from Kiev on the 1 May (day 5), whilst column eight shows that found on swabs from the deck of a ship which was in the Baltic at the time of the accident (5). In both cases, the fission product distributions are fairly similar to that of the original fuel. The Kiev shoe shows an enhancement of some of the actinides over what is expected for the average fuel burn-up, but this may indicate that the activity originated from a region of the core with higher than average burn-up. The only Soviet sampling data including both fission products and actinides relate to soil measurements (Ref 1, Table 4.6). After being reduced to Cs137 RRFs, they have been included in column one, of Table 6, and show a marked depletion in actinides.

Extensive measurements of the radionuclides composition in air and ground samples were made throughout eastern and western Europe during the period following the accident. These show a different pattern from the measurements on the closer-in samples given in Table 6. Air sampling data collected in Finland, Sweden, Denmark, the Netherlands, Hungary, and the UK reveal a remarkably consistent radionuclide distribution. Averaged results of such samples are shown in Table 7. Compared with Table 6, there is a marked depletion in the relative releases of the non-volatile elements, the composition being dominated by caesium, iodine and tellurium. There is also a small but definite enhancement in the contribution of ruthenium and barium relative to lanthanum, cerium and the actinides. The samples on which Table 7 is based have travelled a great distance, and clearly do not represent what came out of the reactor. The apparent depletion of iodine relative to caesium is probably an artifact of the sampling methods employed. Most of the air sampling used filters which collected only particulate-borne activity. However, in a few locations, the particle filters were backed up by charcoal traps to trap any gaseous iodine present. In all locations where this was done, some 70-80% of the iodine was collected on the charcoal. It thus appears that the low iodine fraction shown in Table 7 is the result of incomplete trapping of iodine at most sampling stations. This view is reinforced by comparison with isotope-specific ground samples, which show iodine RRFs generally close to unity. Thus, caesium and iodine were released to the same fractional extents in that part of the source which reached western Europe, but iodine was transported in both particulate and gaseous forms.

In summary, it is clear that a distinct geographical fractionation of the material released from the Chernobyl reactor occurred. It is now established that the activity measured in western Europe during the week or so following the initiation of the accident was released early, during Stage 1 and the first day or two of Stage 2. Only very fine aerosol-sized particles or gases could be transported over the large distances to reach western Europe, and the proportion of the early release in these forms was highly enriched in the volatile fission products, and contained little fuel-based material. However, a larger-sized fraction was also released which settled out more rapidly closer to the reactor, and this was composed mainly of fuel-based material enriched to a lesser extent in the volatile fission products.

6.4 Absolute Activities Released

We finally consider the total releases of radioactivity in the Chernobyl source term. Firstly, the Soviet estimates relating to that part of the activity which settled out within the boundaries of the USSR are examined, and then the information on the activity transported to the rest of the world is reviewed.

The Soviet information to the 1986 Vienna conference (1) contained, in addition to the total daily activity discharges shown in Fig 1, a detailed breakdown of the releases for a sub-set of radionuclides. The latter gives the actual activity releases for these nuclides on day 0 (26 April), and the cumulative releases at the end of the discharge on the 6 May, corrected for radioactive decay to that date. These Soviet data are reproduced in columns 1 and 5 of Table 8. The corresponding release fractions and relative release fractions based on the FISPIN core inventories are also shown. Thus, column 3 shows that on the 26 April, some 3-5% of the volatile fission products and 0.3 - 0.7% of the non-volatile nuclides were released from the core. However, by the 6 May the cumulative release fractions (column 7) were 10-20% for the volatile species against 2-5% for the non-volatile, so that the proportion of non-volatile material in the overall discharge was higher than on day 0. These distributions between volatile and non-volatile constituents are in line with the Soviet sampling data discussed above. Indeed, ground sampling data acquired in the days and weeks after the accident started appear to have provided the basis of the quantitative release estimates reproduced in Table 8. The only additional information from the Soviet Union since August 1986, on the size of the release is given in reference 2. This reconfirms earlier estimates of the cumulative release of Cs137 as about 0.8 MCi (0.03 EBq).

The Soviet estimate for the total released activity (excluding noble gases) present in the environment on the 6 May is 50 MCi (1.85 EBq), ± 50%. This represents the sum of the daily discharges, corrected for radioactive decay to the 6 May, shown in Fig 1. In principle, these decay-corrected data can be converted back to actual daily releases by multiplying by suitable decay factors. For example, it is readily calculated that the sub-set of radionuclides listed by the Soviets and reproduced in Table 8, released with the composition shown there for day 0, would have undergone a decay in total activity by a factor of 2.8 by the 6 May. The whole core would have undergone decay by a factor of 3.6 over this period. Thus, since 12 MCi remained on the 6 May, the actual release on the 26 April should have been in the range 34 - 43 MCi. However, a footnote in the Soviet information on the daily releases quotes the actual day 0 release as 20 - 22 MCi. This latter value is inconsistent with the above treatment, and an additional argument suggests that it must be in error. The total activity associated with the sub-set of radionuclides listed by the Soviets for the day 0 release is 15 MCi when the noble gases are excluded. If, now, allowance is made for additional isotopes of the listed elements which contribute significantly to the radiological consequences and must have had similar release fractions to the lead members, a total activity release of 41 MCi can be estimated for the 26 April. This figure is within the range calculated above, and supports the view that the actual release on day 0 was around 40 MCi (1.5 EBq), twice the value quoted by the Soviets. Considerable uncertainties must thus be attached to the estimates of total activity releases at Chernobyl, but an attempt is made in Table 9 to put together a coherent account based on the Soviet information.

No allowance was made in the Soviet estimates for activity transported across the borders of the USSR. There is now clear evidence that significant fractions of the iodine and caesium released were transported into eastern and western Europe, and further afield. A survey of the radiological impact of the accident on OECD countries (6) provides detailed information on the deposition of I131 and Cs134 + Cs137 from which quantitative estimates of the activities of these isotopes which settled out in various geographical regions can be made. The integral depositions of I131 and Cs134 + 137 on the OECD countries of western Europe amount to 0.097 EBq and 0.019 EBq respectively. No reference date is available for the I131 measurements, but assuming that the 6 May is appropriate, the corresponding contribution to the release fraction from the Chernobyl core is 7.6%. For the caesium isotopes, an average release fraction contribution of 5.4% is derived with less uncertainty, since the radioactive decay effects are negligible. Interpolating the available data to estimate the deposition on the sea areas of western Europe (Baltic, North Sea, Adriatic) leads to further release fractions contributions of 2.7% for I131 and 2.4% for Cs134 + 137. Data on the deposition of these isotopes in eastern European countries are much less complete, but a rough extrapolation from the western European data indicates that about 6% of the I131 and 5% of the Cs134 + 137 in the Chernobyl core were deposited in the eastern European countries bordering the USSR. Thus, in total, it appears that about 16% of the I131 and 11% of the Cs isotopes in the Chernobyl inventory were transported across the Soviet border and deposited in the rest of Europe following release in the accident. Since deposited activity was measured as far afield as the USA, Canada and Japan, further allowance should be made for this in trying to assess to world-wide total of radionuclides which escaped from the USSR. However, integral estimates of the more remote contributions are too uncertain to be worthwhile. A summary of the release fraction contributions discussed above is given in Table 10. It can be concluded that the total release fraction for I131 should be increased from the 20% quoted by the Soviets to at least 36%, and those for the caesium isotopes from around 15% to at least 26%. Similar estimates for an increased release fraction of caesium have been made elsewhere (7,8).

The revised estimates of the Chernobyl release fractions are included in Table 1 for comparison with source terms calculated in the Sizewell B Safety Case. The pattern of fractions is similar to UK-5, but with rather more lanthanides and actinides. It must be emphasised that the pattern of release in time and other features of the Chernobyl source term are quite different. UK-5 would be a single puff at a notional containment failure time of 4 hours while the Chernobyl release spanned ten days, with a double peaked activity versus time profile.

6.5 Conclusions

The considerable details of the Chernobyl source term provided by the USSR at IAEA meetings (1,2), supplemented by extensive sampling measurements of the released radioactivity in Europe and elsewhere, enable a fairly complete and consistent description of the source term to be developed. There was a prompt release of activity in the initiating event on the 26 April, followed by steadily decreasing releases over the following five days. A large amount of energy was associated with the early release, resulting in a high plume rise and wide dispersion of the released activity. An assessment of the total activity released on the 26 April suggests that this amounted to around 40 MCi, rather than the 20 MCi quoted by the Soviets. Dumping of filtering and shielding materials on the core debris was initially successful in reducing the rate of activity release, but from the 2 to the 5 May, the rate began to rise again as the decay heat steadily drove the debris temperature upwards. On the 6 May, effective cooling of the debris was finally established, and the release terminated abruptly. The radionuclide composition of the released activity exhibits a marked geographical variation. Soviet sampling in a region extending a few tens of kilometres around the reactor shows essentially fuel-based material enriched by factors from 3 to 10 in the volatile fission products iodine, caesium and tellurium. Overall release fractions of 10-20% for the volatile fission products and 3-4% for the non-volatile radionuclides are derived from these Soviet samples. Additionally, significant quantities of the released iodine and caesium were transported across the Soviet border, and samples measured there show much greater enrichment in these nuclides compared with the Soviet samples. It is estimated that the release fractions of iodine and caesium should be increased by at least 16% and 11% respectively above the Soviet values to allow for this remotely-deposited activity. The total release fractions for iodine and caesium then approach those characteristic of the worst type of severe nuclear reactor accident which has been contemplated in the West.

6.6 References

1 USSR State Committee on the Utilisation of Atomic Energy. 'The Accident at the Chernobyl Nuclear Power Plant and its Consequences'. Information compiled for the IAEA Experts' Meeting, 25-29 August 1986, Vienna. Part II, Annex 4.

2 V G Asmolov and others. The Chernobyl Nuclear Power Station Accident: One Year Afterwards. IAEA Conference on Nuclear Power Performance and Safety, 28 September - 2 October 1987, Vienna. IAEA-CN-48/63.

3 R F Burstall, A Tobias. Private communications.

4 An Assessment of Accident Risks in US Commercial Nuclear Power Plants. Reactor Safety Study (WASH-1400). USNRC 1975.

5 J P Longworth and A Tobias. Use of Activity Measurements in the Plume from Chernobyl to Deduce Fuel State Before, During and After the Accident. CEGB Report TPRD/B/0844/R86 (July 1986).

6 OECD/NEA. The Radiological Impact of the Chernobyl Accident in the OECD Countries. OECD, Paris, 1987.

7 R S Cambray et al. Observations on Radioactivity from the Chernobyl Accident. Nuclear Energy, 26, 77 (1987).

8 A Aarkrog. The Radiological Impact of the Chernobyl Debris compared with that from Nuclear Weapons Fallout. J. Environ. Radioactivity, 6, 151 (1988).

TABLE 1

Fractions of Radioisotopes Calculated to be Released ('Source Terms')

Category	Equivalent Crude Category	Xe-Kr	I*	Cs-Rb	Te	Ba-Sr	Ru	La
UK-1	A	0.9	0.5	0.5	0.3	0.06	0.02	0.004
UK-5	B	1.0	0.3	0.3	0.5	0.04	0.03	0.006
UK-10	C	6E-3	1E-5	1E-5	2E-5	1E-6	1E-6	2E-7
UK-12	D	5E-2	1E-6	1E-6	2E-7	1E-7	4E-8	4E-9
Chernobyl		1.0	0.4	0.25	>0.1	0.04	0.05	0.03

* Revised First Estimates

TABLE 2

FISPIN Prediction of Chernobyl Inventory - Shutdown Activity in EBq

Nuclide	Activity	Nuclide	Activity
Kr 85	.02	Ru 103	4.37
Kr 85m	.34	Ru 105	1.05
Kr 87	.27	Ru 106	.93
Kr 88	.65	Mo 99	5.53
Xe 133	6.60	Tc 99m	5.09
Xe 135	2.36	Rh 105	2.57
I 131	2.98		
I 132	4.21	La 140	5.95
I 133	5.10	Ce 141	5.56
I 134	1.22	Ce 143	4.42
I 135	2.77	Ce 144	3.88
		Pr 143	5.26
Cs 134	.11	Nd 147	2.16
Cs 136	.08	Y 90	.20
Cs 137	.24	Y 91	4.68
Rb 86	4.87(-3)	Zr 95	5.83
		Zr 97	4.01
Te 127	.24	Nb 95	5.89
Te 127m	.03		
Te 129	.39		
Te 129m	.14	Np 239	47.27
Te 131m	.37	Pu 238	7.59(-4)
Te 132	4.10	Pu 239	9.33(-4)
		Pu 240	1.28(-3)
Ba 140	5.75	Pu 241	1.01(-1)
Sr 89	3.57	Am 241	6.58(-5)
Sr 90	.19	Cm 242	1.15(-2)
Sr 91	2.54	Cm 244	5.48(-5)

Note: 4.87(-3) = 4.87×10^{-3} etc.

TABLE 3

FISPIN Prediction of Chernobyl Inventory -
10 days after Shutdown Activity in EBq

Nuclide	Activity	Nuclide	Activity
Kr 85	.024	Ru 103	3.667
Kr 85m	-	Ru 105	1.99(-4)
Kr 87	1.41(-5)	Ru 106	.916
Kr 88	-	Mo 99	.445
Xe 133	2.06	Tc 99m	.428
Xe 135	1.22(-4)	Rh 105	.024
I 131	1.28		
I 132	.50	La 140	3.84
I 133	.001	Ce 141	4.50
I 134	-	Ce 143	.028
I 135	-	Ce 144	3.79
		Pr 143	3.45
Cs 134	.109	Nd 147	1.15
Cs 136	.048	Y 90	.199
Cs 137	.239	Y 91	4.18
Rb 86	.003	Zr 95	5.24
		Zr 97	4.86(-4)
Te 127	.073	Nb 95	5.83
Te 127m	.036		
Te 129	.076		
Te 129m	.117	Np 239	2.49
Te 131m	.001	Pu 238	7.71(-4)
Te 132	.488	Pu 239	9.45(-4)
		Pu 240	1.28(-3)
Ba 140	3.34	Pu 241	1.01(-1)
Sr 89	3.11	Am 241	7.03(-5)
Sr 90	.199	Cm 242	1.11(-2)
Sr 91	2.76(-4)	Cm 244	5.47(-5)

TABLE 4

Comparison of FISPIN and Soviet Shutdown Inventories

(E Bq)

	FISPIN	Soviet	Soviet/FISPIN
I131	2.98	3.1	1.04
Te132	4.10	2.7	0.66
Cs134	0.11	0.18	1.64
Cs137	0.24	0.28	1.17
Sr89	3.57	2.5	0.70
Sr90	0.19	0.20	1.05
Ba140	5.75	4.9	0.85
Mo99	5.53	6.07*	1.10
Ru103	4.37	4.9	1.12
Ru106	0.93	2.1	2.25
Ce141	5.56	5.5	0.99
Ce144	3.88	3.2	0.82
Zr95	5.83	4.8	0.82
Np239	47.3	26.3	0.56

* Soviet value ÷ 10 to correct assumed misprint.

TABLE 5

Isotope Ratios

	FISPIN* Calculation	Soviet Calculation	"average soil"	airborne day 0	airborne day 6	airborne day 7	airborne day 8	Kiev "student's shoe"	distant measurements
Cs $\frac{134}{137}$	0.46 (0.46)	0.64	0.94	-	0.43	0.43	0.46	0.46	0.5
Ru $\frac{106}{103}$	0.21 (0.25)	0.43	0.27	0.38	0.27	0.43	0.19	0.28	-
Ce $\frac{144}{141}$	0.70 (0.84)	0.58	-	1.14	0.80	-	0.80	-	-

* value at shutdown; day 10 value in brackets.

TABLE 6

Relative Release Fraction - Based on ^{137}Cs

Nuclides	Soil samples (average)	Soil sample* richest in non-volatiles	Airborne* day 0	Airborne* day 3	Airborne+ days 6-9	Airborne day 7	Kiev (student's shoe)	Baltic Ship
Cs 137	1.00	-	-	-	1.00	1.00	1.00	1.00
Cs 134	2.07	-	1.91	1.76	.92	.91	1.00	1.32
I 131	.88	1.00	1.00	1.00	.98	.99	1.20	0.74
Te 132	.99	-	2.55	2.55	1.36	1.57	1.45	1.15
Ba 140	.14	.84	.29	.59	.15	.05	-	1.55
Zr 95	.12	1.19	.36	.40	.24	.00	.82	1.31
Mo 99	-	-	.35	.36	.48	.30	-	-
Ru 103	.16	.08	.26	.24	.26	.12	.98	0.60
Ru 106	.14	-	.45	.46	.44	.21	1.29	-
Ce 141	.13	1.11	.13	.13	.21	.01	-	1.21
Np 239		.78	.25	.05	.14	.005	1.00	
Pu 238	8.95E-3						3.76	
Pu 239+240	6.78E-3						3.00	
Pu 241							1.81	
Am 241							10.17	
Cm 242							4.13	
Cm 244							6.97	

* no Cs 137 data available - normalised to I 131 instead

\+ average over 4 days

TABLE 7

Relative Release Fractions from Chernobyl based on Measurements at Distant Sites

Element	Isotope	$t_{\frac{1}{2}}$	Relative Release Fraction
Cs	137	30y	1.0
	134	2.06y	
I	131	8.04d	0.35-0.50
	132	2.30h	
	133	20.8h	
Te	132	78.2h	0.55-0.90
Ba	140	12.7d	0.035-0.050
La	140	40.3h	<0.005**
Ru	103	39.4d	0.055-0.120
	106	368d	
Ce	141	32.5d	<0.005
	144	285d	
Nb	95	35.1d	<0.005
Np	239	2.36d	0.002-0.014

** Based on only one measurement which distinguished La140 from Ba140.

TABLE 8

Absolute Releases and Fractions

	Activity out 1st day (EBq)	Inventory+ day 0	RF	RRF	Activity out up to day 10 (EBq)	Inventory+ day 10	RF	RRF
Kr 85m	.006	.34	1.7 E-2	.38	.019*	.025*	.762	4.93
Xe 133	.190	6.60	2.8 E-2	.62	1.670	2.06	.808	5.23
I 131	.170	2.98	5.7 E-2	1.24	.270	1.28	.209	1.35
Cs 134	.005	.109	5.1 E-2	1.11	.019	.109	.174	1.12
Cs 137	.011	.239	4.5 E-2	1.00	.037	.239	.154	1.00
Te 132	.150	4.10	3.6 E-2	.79	.048	.48	.098	.63
Ba 140	.014	5.75	2.4 E-3	.05	.160	3.34	.047	.30
Sr 89	.009	3.57	2.5 E-3	.05	.081	3.11	.025	.16
Sr 90	6E-4	.19	3 E-3	.06	.008	.19	.040	.25
Ru 103	.022	4.37	5 E-3	.10	.120	3.66	.032	.21
Ru 106	.007	.93	7.4 E-3	.16	.059	.91	.064	.41
Mo 99	.017	5.53	3 E-3	.06	.110	.44	.247	1.59
Ce 141	.015	5.56	2.6 E-3	.05	.100	4.50	.022	.14
Ce 144	.017	3.88	4.3 E-3	.09	.089	3.79	.023	.15
Zr 95	.017	5.83	2.9 E-3	.06	.140	5.24	.026	.17
Np 239	.100	47.27	2.1 E-3	.04	.044	2.49	.017	.11
Pu 238	3.7 E-6	7.59 E-4	4.8 E-3	1.06 E-1	3 E-5	7.71 E-4	.038	.25
Pu 239	3.7 E-6	9.33 E-4	3.9 E-3	8.63 E-2	2.6 E-5	9.45 E-4	.027	.17
Pu 240	7.4 E-6	1.22 E-3	6 E-3	1.31 E-1	3.7 E-5	1.28 E-3	.028	.18
Pu 241	7.4 E-6	1.01 E-1	7.3 E-5	1.59 E-3	5.2 E-3	1.01 E-1	.051	.33
Pu 242	1.1 E-8	8 E-7	1.3 E-2	2.99 E-1	7.4 E-8	8 E-7	.092	.59
Cm 242	1.1 E-4	1.15 E-2	9.5 E-3	2.07 E-1	7.8 E-4	1.11 E-2	.070	.45

* Kr 85 not Kr 85m

+ FISPIN

TABLE 9

Total Activities Released Excluding Noble Gases in MCi (EBq)

	Corrected to	Total	Subset of nuclides on Table 8
On day 0	day 0	40 (1.5)*	15 (0.56)
Up to day 10	day 10	50 (1.9)	35 (1.3)
Up to day 10	day 0	120 (4.4)*	-
Total inventory	day 0	5750 (213)	-
Total inventory	day 10	1630 (60)	-

* Calculated using FISPIN Decay Curve

TABLE 10

Fractions of Chernobyl Core Inventory of Iodine and Caesium deposited outside the USSR

Location	Fraction of Inventory, %	
	I131	Cs134+137
W.European land (1)	7.6	5.4
E.European land (2)	6	5
European sea (2)	2.7	2.4
USA (1)	0.1	0.1
Canada (1)	0.08	0.1
Japan (1)	0.03	0.01

Notes

(1) Based on reference 6

(2) Estimated by extrapolation/interpolation

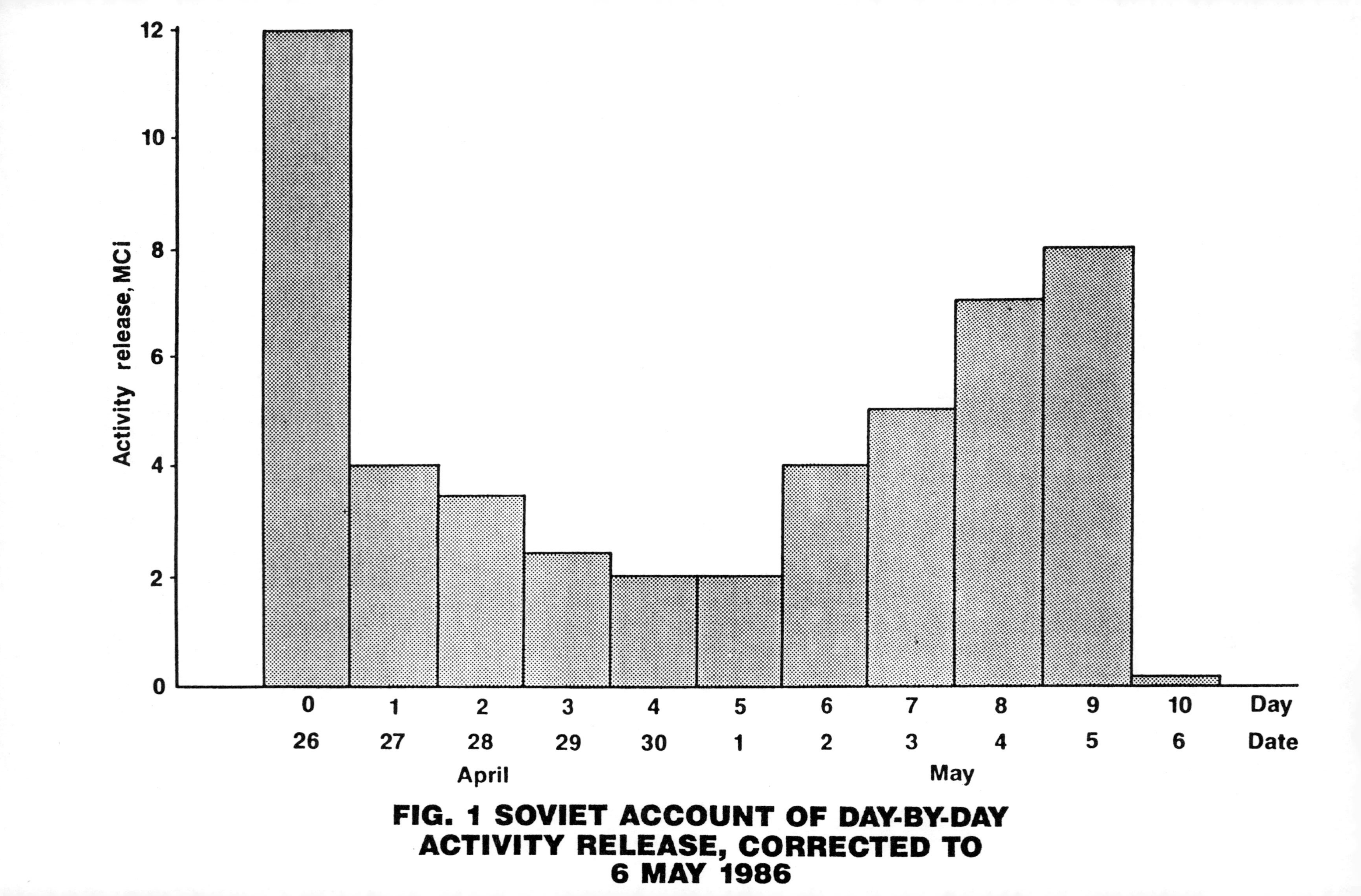

FIG. 1 SOVIET ACCOUNT OF DAY-BY-DAY ACTIVITY RELEASE, CORRECTED TO 6 MAY 1986

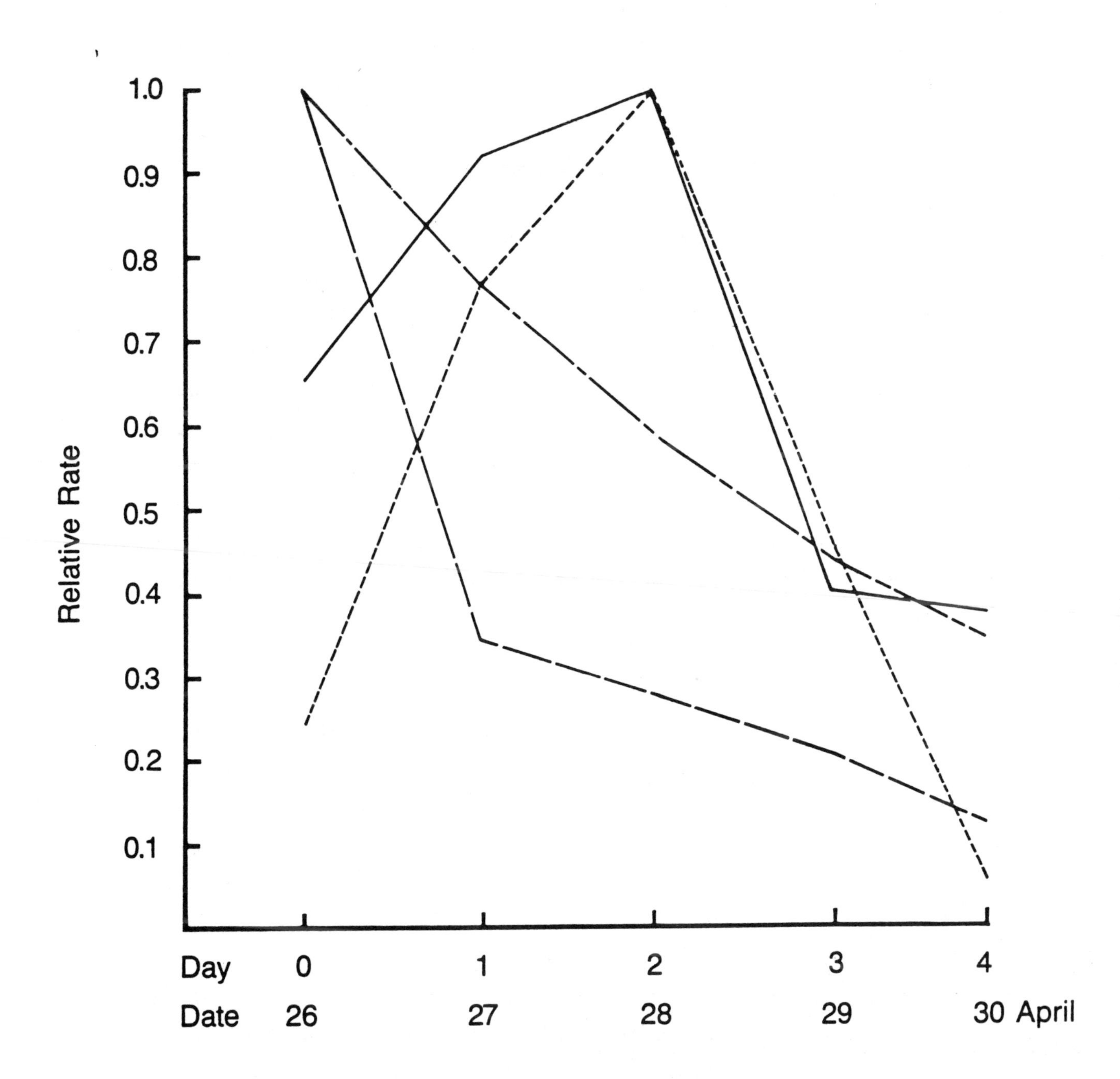

FIG. 2 SUMMARY OF SOVIET DATA ON THE RELATIVE ACTIVITY RELEASE AND DEPOSITION RATES, DAYS 0 - 4

SECTION 7: ENVIRONMENTAL CONSEQUENCES

Following the accident at Chernobyl, radioactive material was swept across Europe resulting in increased dose levels. This section considers the time dependent pattern of the spread of contamination throughout Europe and presents estimates of the collective dose to various countries. At the outset, it should be appreciated that the assessment is based upon data available during the year following the accident; certain improvements might arise on the strength of further information. Most of the work described here has been published elsewhere[1].

7.1 Atmospheric Dispersion Across Europe of Material Released from Chernobyl

Increased activity levels were first reported on 28 April from environmental monitoring stations in Finland and Sweden, where external dose rates in certain locations exceeded normal background levels by a factor of ten or more. On succeeding days elevated radioactivity concentrations were detected throughout Europe until almost complete coverage had been achieved by 3rd May. Based upon reported measurements conveyed through international bodies (IAEA, WHO and NEA), complemented by computer calculations, it has been possible to assemble a picture of the pattern of dispersion of the material released from the core of the damaged reactor, as it affected western Europe. The progression of this pattern with time is illustrated in Figures 1 to 6. The figures indicate how the external exposure rate varied across Europe from 28 April to 3 May. Note that the monitoring data used as a basis for these plots exhibits a marked patchiness, deriving from (patchy) rainfall patterns and, possibly, uncertainty in the environmental measurements. Such large variations over relatively short distances are not shown in the figures, so that they provide a general picture of the spread of the contamination across Europe. Note also that the figures may be subject to some slight revision as more monitoring information becomes available.

Referring to Figures 1 to 6, it can be seen that by the 28 April (Figure 1), radiation dose levels had increased in Scandinavia (as has already been noted above), resulting from the generally north-westerly trajectories prevailing at that time. By the 29 April (Figure 2), the contamination had spread further across the Scandinavian countries, with lateral dispersion increasing the width of the plume. On the 30 April (Figure 3), central Europe was beginning to be affected, reflecting a trajectory which was initially north-westerly, but which subsequently veered westwards in the vicinity of the Baltic sea. (At this point it should be appreciated that Figures 1 to 6 are based upon measured dose rates, so that although by the 30 April the plume of material was largely over central Europe, dose levels were still relatively high in Scandinavia, reflecting earlier deposition of material from the plume). From the 1 to 3 of May (Figures 4 to 6), the plume spread to the west, north and south, essentially covering Europe. This further spreading of the plume was influenced by an anticyclone which was moving eastwards across central Europe. Indeed the contamination of the UK resulted from air being convected northwards behind the area of high pressure; the higher contamination in the north of the UK reflects greater rainfall rates during plume passage.

As far as Europe is concerned, from the 3 May the dose levels generally stabilised and fell. One of the more notable exceptions to this is Scandinavia, where increased air concentrations were observed around 8 May. This almost certainly represents material discharged during the latter part of the Chernobyl release (around 5-6 May).

Before leaving this discussion it is worth noting the similarities between the contamination patterns of Figures 1 to 6 and those generated by long-range trajectory models. Many such studies have been performed (see bibliography); the interested reader is referred to the work of, for example, ApSimon et al[2]. Additionally, it is of interest to note the more detailed studies of UK activity levels performed by Clark and Smith[3].

In addition to the very wide dispersion brought about by the changing meteorology over the several days during which emissions from the damaged plant took place, it seems likely that, initially, material was distributed over a considerable range of elevation. Material transported at very high altitudes ($\geq$ 1km) may have been responsible for the subsequent observations of elevated activity levels in countries bordering the Pacific Ocean.

7.2 Dosimetric Assessment for Western Europe

An estimate of the dosimetric impact on Western Europe from Chernobyl may be obtained by utilising the monitoring data collected and published by the various national agencies responsible for radiological protection. Such an assessment is presented here, although it is important to recognise that conclusions are necessarily drawn from preliminary data obtained during the year following the accident.

At the outset it should be appreciated that, within any country of Western Europe, there is some variability in the measured concentrations of radioactivity, arising from the complicated patterns of atmospheric dispersion and rainfall during passage of the plume. Clearly any dosimetric assessment needs to take account of this distribution in relation to the distribution of population. Presented here are estimates of the mean (population-weighted) individual dose for various countries, based on estimated mean environmental concentrations. The (mean) dose estimates are, therefore, subject to a degree of uncertainty, up to around a factor of a few depending on the country, arising solely from this averaging process. Other sources of uncertainty are discussed below.

Dosimetric pathways contributing to radiation exposure include the inhalation of activity during passage of the plume, ingestion of contaminated foodstuffs and external irradiation from deposited activity. In addition to these, β-dose to the skin and external exposure to radiation from the passing cloud of activity can also contribute to total dose levels; however, these mechanisms are transient in nature and have been shown to make up only a very small fraction of the total effective dose to a representative individual[4]. Each of the pathways considered in the present analysis is discussed briefly below.

7.2.1 Inhalation pathway

In general, significant elevated concentrations of activity in air were present for a few days and direct measurements of ambient levels of the radiologically important nuclides (isotopes of Cs, I and Ru) were monitored. This data allows time integrated air concentrations to be estimated. Used in conjunction with standard values of inhalation rate and dose per unit intake for adults[5], a mean, individual, committed effective dose equivalent can be calculated for the inhalation pathway.

7.2.2 Ingestion Pathway

By contrast with inhalation exposure, radioactivity transferred to man through incorporation in foodchains is available over a more extended period, so that monitoring data is unlikely to give a full picture of the average intake of activity via foods. It is therefore necessary to turn to mathematical models representing the temporal pattern of appearance of radionuclides in different foods following an initial deposit. For milk and green vegetables, doses are estimated from measured concentrations of Iodine and Caesium activity in these foodstuffs on the basis of models used by NRPB in deriving reference levels for the introduction of countermeasures affecting food[5]. Consistent with the approach adopted in these models, individual doses are calculated from the peak values of the distributions of activity concentration with time and are integrated to 50 years. However, activity losses occurring in preparation of vegetables for consumption are neglected here for the sake of conservatism. (Nevertheless, it is notable that there is some evidence suggesting that Chernobyl fallout was not always easily removed during processing). Representative average concentrations of activity are used for each country, while average consumption rates typical of the UK adult population[6] are assumed in order to characterise intake. In some cases, the absence of direct monitoring data for foodstuffs has necessitated the estimation of peak concentrations in milk and green vegetables from measured deposition levels.

One advantage of using measured levels of contamination in foodstuffs with the models is that the effect of certain countermeasures is implicitly included; thus, if dairy cattle were removed from contaminated land or kept indoors for a period of time (eg in Scandinavia and the Low Countries), measured activity levels in milk will be low and reflected in assessed doses. While there is some evidence that other, mainly voluntary, countermeasures were introduced affecting normal food distribution and consumption patterns throughout Europe, these have been ignored in the dosimetric assessment. It might therefore be expected that the results of the calculations yield a slightly conservative estimate of the overall dosimetric impact.

In addition to the contribution to ingestion doses from milk and green vegetables, estimates are made of the intake of Caesium in supplies of beef and lamb. Here committed doses from ingestion are derived from reported ground contamination levels, using published data on the transfer of activity through foodstuffs(7). Again, typical UK adult consumption rates are assumed(6), with appropriate values for the committed dose equivalent per unit intake(8). Calculations are extended to include contributions to ingested activity arising in meat over the next 50 years.

Scoping estimates based upon the same suite of models as those used above(7) suggest that, by comparison, the contribution to mean individual committed doses from other foods (root vegetables, cereals etc) will be relatively small (well within bounds of uncertainty). This is due in part to the recorded absence of significant quantities of the radioisotopes of Strontium in environmental monitoring (usually considered to be important for foods contaminated by uptake from soil) and to the delay of a number of months between the accident and the cereal harvest. Ingestion dose calculations are therefore limited to contributions from milk, green vegetables and meat.

7.2.3 External exposure

External exposure to activity deposited within each country during passage of the plume will continue for several decades. During this time, the predominant contribution to external doses will be due to the decay of radioisotopes of Caesium. For the present assessment, the population-weighted average ground concentrations of these nuclides are used in conjunction with appropriate dose conversion factors taking account of decay and migration into the soil(5). The dose calculations assume a shielding factor of 0.36 for protection by buildings and involve integration over a period of 50 years following the accident.

7.2.4 Note on calculations for UK

A small additional degree of sophistication is introduced into estimates of mean individual dose in the UK. It is well known that a fairly sharp division exists between contamination levels in the northern and north-western parts of the United Kingdom and the remainder of the country, due to different rainfall patterns at the time the plume was passing. Dosimetric calculations are therefore made separately, assuming average contamination levels characteristic of the two regions. A population weighting factor of 18% is applied to doses calculated for the 'north' and 82% to those estimated for the 'south'. The only exception to this method for averaging doses is applied in the case of lamb consumption, since a large proportion of the country's sheep farming is in the more heavily contaminated region. Indeed restrictions on sale and slaughter of lambs have been in force in certain of the worst affected regions. Mean individual doses to members of the UK population from consumption of lamb are therefore determined from the characteristic average ground concentrations of Caesium reported for the 'north' only, taking into account that the UK is only 77% self-sufficient in mutton and lamb. (Note that the average concentration of Caesium in the more heavily contaminated region is less than that in those individual areas where restrictions have been applied. Potential reductions in population dose achieved through the imposition of restrictions are therefore not incorporated in the assessment).

7.2.5 Results

Table 1 shows the results of a dosimetric analysis for Western Europe, in terms of the contributions to the total collective effective dose (integrated to 50 years) from the various pathways considered. It can be seen that the contribution from inhalation is less than 10%, while those from external irradiation and ingestion are roughly similar. The total collective dose to Western Europe is estimated to be approximately 76,000 man Sv.

Table 2 shows the estimated distribution of dose among West European countries; for each country the mean individual effective dose integrated to 50 years and the total 50 year collective effective dose is presented. In some cases, the contributions to total dose from each of the pathways differ somewhat from the general picture of Table 1; comments on such deviations and other considerations are included, where necessary, in Table 2. Of course, many of these countries have published their own estimates of dose (see bibliography). The methods used by each nation differ in their degree of sophistication and the time periods over which dose is considered to be accumulated. However, in general, these estimates compare favourably with those calculated here, bearing in mind likely uncertainty levels (see comment in Section 7.4).

As noted above, the total collective dose for Western Europe is estimated to be approximately 76,000 man Sv. In order to gauge the potential implications of this level of exposure for the health of the population, it is necessary to make assumptions about the relationship between radiation dose and risk. The results of several new studies and reviews concerning this relationship are expected to become available in the next few years. In the main, this work concerns radiation exposure arising from the Hiroshima and Nagasaki atomic bombs, where individual doses were some orders of magnitude higher than those predicted here as a consequence of the Chernobyl accident. A formal recommendation by the International Commission on Radiological Protection (ICRP) on the appropriate dose-risk relationship awaits the conclusions of the ongoing work; however, it seems likely that the new data could suggest a revision to existing risk estimates, raising the total fatal cancer risk by a factor of order 2[9]. Using a linear relationship between dose and cancer mortality risk (generally considered to be conservative for doses received at or below background exposure levels) to obtain risk estimates corresponding to the relatively low doses from Chernobyl, this would imply a risk coefficient of perhaps around 0.03 cancer fatalities per man Sv. This, in turn, implies a total number of cancer fatalities in Western Europe, arising from the Chernobyl accident over the next decades, of around 2300. Note, however, that, due to the conservatisms involved, this figure should be regarded as a likely upper estimate.

A similar assessment to that above has been performed by the UK National Radiological Protection Board (NRPB) for the countries of the European Community[10]. For these countries the results obtained were similar (within reasonable bounds of uncertainty) to those presented here. In addition, the NRPB have undertaken a slightly more detailed analysis for the UK than that considered above[11]. This divided the country into several regions depending on the recorded distribution of activity levels.

However, the resulting average dose for the population as a whole is very similar to that calculated here.

7.3 Dosimetric Assessment for Eastern Europe

Information on levels of radioactive contamination in Eastern Europe is relatively limited and an assessment of the dosimetric impact for these countries is therefore subject to considerable uncertainty. Here, calculation of mean individual dose is performed by scaling from the values previously estimated for countries in Western Europe according to the ratio of activity concentrations in deposited material or foodstuffs, whichever data are available.

Table 3 summarises mean individual doses and total collective dose commitments estimated in this way for Eastern European countries excluding the Soviet Union. The total collective dose is estimated to be approximately 100,000 man Sv which, using a linear dose risk relationship in the manner described above for the dose received in Western Europe, corresponds to a total of around 3,000 fatal cancers. Note that this figure should be viewed in the context of the assumptions made regarding the relationship between dose and risk, as discussed earlier.

7.4 General comment on estimated doses

It must be appreciated that the dose estimates in Tables 1, 2 and 3 are subject to some uncertainty. Firstly, as already noted above, the use of a weighted contamination level for each country studied may give rise to uncertainty levels up to around a factor of a few. Secondly, the use of monitoring data (with its associated uncertainties), and the application of standard dosimetric models using UK consumption data may give rise to similar degrees of uncertainty. This should be borne in mind when considering the data in the Tables and when comparing the doses with those estimated by others. It is worth noting that whole body measurements of uptake by members of the public, including those in the United Kingdom[12], suggest that the models widely used to evaluate the ingestion of radio-isotopes of caesium tend to overestimate the dose from these nuclides. The uncertainty in the overall dose calculations is clearly greater for Eastern as compared to Western Europe; indeed the former should be regarded as order of magnitude estimates.

The preliminary nature of the above dose calculations should also be noted; they have been performed using data available during 1986-87 and may be subject to some revision as more information is obtained.

7.5 Consequences in the USSR

The following is a discussion of some of the impacts in the USSR resulting from the accident, based primarily on the presentations made by the Soviet delegation to the IAEA in Vienna[13,14].

7.5.1 Site emergency and medical response

At the reactor site, an immediate priority following the accident was attached to fighting a number of fires which had broken out in and around the reactor building and turbine hall, threatening the safety of the station's third reactor unit. Within an hour of the accident the fire team stationed at the plant, together with firemen from the nearby towns of Pripyat and Chernobyl, were beginning to bring the worst of these fires under control. It appears that there was some difficulty initially in accurately reporting the severity of the situation to personnel at the plant and to the relevant authorities in Moscow. Nevertheless, site emergency response initiated within the first two hours included arrangements for beds to be made available in local hospitals and the issue of stable iodine tablets to plant personnel. Furthermore, a specialist team had been called out and flown from Moscow within ten hours of the accident.

From among the on-site personnel, some 300 required hospital treatment. These included reactor and electrical plant operating personnel, site emergency squads and, particularly, members of the fire brigades. Of those examined, 203 were diagnosed as suffering from acute radiation syndrome arising from absorbed doses in the range 1-16 Gy (almost entirely from external radiation, in particular from debris on the site and airborne contamination). Two days after the accident, 129 casualties were flown to Moscow for specialist treatment and the remainder taken to Kiev. Despite efforts to provide bone marrow transplants for the worst affected victims (diagnosed irreversible depression of bone marrow function), only limited success was achieved. All but one of 13 patients receiving this support died. A total of 29 fatalities has been reported from among those hospitalised and diagnosed as suffering from acute radiation effects. Two additional deaths are reported to have occurred in the immediate aftermath of the accident. Based on prior understanding of human biological response to very high acute radiation exposures, a lethality substantially higher than that which was observed might have been expected, in the absence of therapeutic treatment, for the range of doses experienced by the casualties among the site personnel. The considerable medical care which was made available within a relatively short time, including blood transfusions, chemotherapy and antibiotic administration, together with techniques to prevent infection, appears to have been generally effective (within the limits imposed by the severity of injuries) in achieving an increased survival rate.

It has been claimed in news reports that three additional deaths occurred during decontamination work, although it is not clear whether these were due to radiation exposure or were related to normal industrial accidents.

7.5.2 Public health measures

It has been emphasised in the literature presented by the Soviet authorities that no individual member of the population away from the reactor site itself incurred a radiation dose above the threshold for clinically manifest symptoms of acute radiation syndrome. An emergency control centre, established in the town of Chernobyl with the support of State Committee for Atomic Energy and the specialist team from Moscow, directed the emergency response as it affected the population surrounding the reactor site. Precautionary evacuation of the town of Pripyat, less than 10 km from the site, was viewed as an immediate objective. However, the initial plume of radioactive material released from the damaged reactor had missed the town but contaminated evacuation routes laid down under existing emergency plans. During the morning of 26 April, the day of the accident, people were instructed to shelter indoors with windows and doors shut; schools and kindergartens were closed. Later that day stable iodine tablets were distributed by volunteers from house to house in the town. During this time, ad hoc evacuation plans were being devised so that by the next day, when radiation levels in the town had begun to rise sharply, the necessary resources (transportation, relocation centres, medical teams) had been organised. The population of 45,000 people were evacuated in 2½ hours, commencing at 2.0pm on 27 April. Individual exposure appears to have been kept below the upper dose intervention level of 750 mSv operated by the Soviet authorities as a reference for emergency planning. During the next few days, the emergency control centre supervised the gradual evacuation of a further 90,000 people from within a radius of about 30 km around the plant as the levels of contamination became more widespread. Substantial medical resources were deployed to monitor the total of 135,000 evacuees, many of whom were found to be wearing contaminated clothing requiring replacement. Nevertheless, no cases of acute radiation syndrome were diagnosed from among the evacuees and, correspondingly, none were hospitalized as a result of radiation-induced effects.

It is planned that an epidemiological survey will be carried out, based primarily upon the 135,000 evacuees from around the Chernobyl site. However, additional members of the population living outside the 30km zone are also to be included since, in some cases, they received greater doses than the evacuees. The scale of the survey will clearly be enormous (the number of people involved is greater than that in the Hiroshima and Nagasaki survivor survey), and will need to continue well into the next century.

A very large medical and public health programme was also established in the months following the accident. In addition to those evacuated from the controlled 30km zone, many hundreds of thousands of people have been subject to medical examination, a large proportion of whom have undergone dosimetric monitoring. This represents a substantial increase above the normal practice. Furthermore, stable iodine was distributed over a wide area to millions of people. The medical programme has been supplemented by a health education campaign and the organisation of special summer camps for children and pregnant women from large cities such as Kiev. The Soviet authorities claim important benefits for public health arising from the programme.

7.5.3 Monitoring and agricultural measures

The emergency control centre was responsible for a variety of other aspects of the emergency response, including radiological monitoring both inside and outside the 30 km evacuation zone, food and water bans, control of the movement of livestock and decontamination. The effort appears to have been supported by the deployment of thousands of troops within the affected area, together with technical teams assembled from appropriate institutions throughout the Soviet Union.

Initially, monitoring operations were directed towards the resolution of essential short-term problems such as evacuation. In the succeeding months, various Soviet state committees and government ministries co-operated in establishing a continuous monitoring system, supplemented by scientific programmes aimed towards improved detection techniques and understanding of the migration of deposited radionuclides. The information gained from this exercise has been used primarily in formulating decisions on long-term measures in the protection of water resources and agriculture, together with decontamination. Results of the monitoring operation have also formed the basis of estimates of the dosimetric consequences for the population of the Soviet Union (see 7.5.4 below).

A significant effort has been devoted to protective actions reflecting the fact that the immediate contaminated area is in the flood plain of several large rivers (notably the rivers Pripyat and Dnieper). These drain into the Kiev reservoir and others on the Dnieper cascade, which are large and important fresh water bodies. Parts of the region are subject to regular annual flooding, providing a mechanism for the ready transfer of contaminated silts and surface waters from the land to rivers. A large-scale construction programme of earthworks and dams was therefore initiated to contain the deposited activity. The result seems to have been largely successful inasmuch as, although activity has been measured since July 1986 on suspended particulate and in solution in water destined for human consumption, contamination of supplies has not reached levels judged unacceptable by the radiation protection standards employed in the USSR.

Both livestock and arable farming had been carried out in the contaminated region prior to the accident. More than 50,000 head of cattle and 9,000 pigs were removed from the 30km zone in the initial period after the release, while access controls were established for personnel and vehicles moving within different areas of the zone.

Overall, approximately 50% of the contaminated land was occupied by agriculture (nearer 25% in the region closer to the reactor site), the remainder generally being uncultivated marshland or forest. Depending upon the degree of contamination in the affected areas, a three-tier decision process was adopted, on the basis of concentrations of Caesium-137. At the uppermost level, agricultural use of land is prohibited, although it is possible that such areas may be transferred to forestry in order to retain some economic value. At intermediate contamination levels, agricultural production has been continued under specific limitations on land use (ie the particular crop or livestock raised) and management. Careful choice of the type of farming can limit to a minimum the transfer of radionuclides through the foodchain to man. Recommended chemical treatments of contaminated soil in these areas include the application of sorbents such as clay suspensions, to fix the radioactivity in a form inaccessible to plant roots, together with the use of lime and mineral fertilizers rich in Calcium and Potassium, thereby reducing the uptake of the chemically similar radioisotopes of Strontium and Caesium. Elsewhere, agricultural practice has been allowed to continue substantially unaltered, except that produce from these and other areas is subject to monitoring and control.

Depending on the level of radioactive contamination, food products have been either (i) interdicted and prevented from any use at all, (ii) redirected to alternative processes or longer storage prior to use or (iii) distributed subject to activity concentrations meeting radiological protection standards. In following these countermeasures, the Soviet authorities appear to have attempted to maximise the economic use of contaminated land while maintaining activity concentrations in foods reaching man at acceptable levels.

7.5.4 Comment on Soviet estimate of collective dose

In the presentation to the IAEA meeting in Vienna in August 1986[13], the Soviet delegation estimated the collective dose in the European part of the USSR from various pathways. For external exposure a collective dose of around 3×10^5 man Sv was reported. For internal exposure resulting from consumption of foodstuffs contaminated with Caesium, they estimated a figure of around 2×10^6 man Sv. Finally, their quoted estimate of the number of thyroid cancer fatalities suggested a collective effective dose from Iodine in milk of around 10^5 man Sv.

These figures are somewhat out of line with previous estimates of collective dose from atmospheric releases of radioactivity including large quantities of Caesium (eg from weapons fallout), which suggest that the contributions from foodchains and external exposure should be roughly equal (see eg reference (15)). Indeed, in subsequent presentations of Soviet collective dose estimates[14], this discrepancy has largely been resolved. The contribution from Caesium ingested in foodstuffs is now estimated to be around an order of magnitude lower than that originally suggested. This reflects the results of an extensive whole body monitoring campaign in the months following the accident, which indicated that the models for Caesium uptake and metabolism used in the original assessment were too pessimistic. In addition to the reduction in dose via ingestion, a revised assessment of the pattern of radioactive decay for the spectrum of fission products released in the accident has led to a slightly smaller estimate of the collective dose from external exposure.

The more recent estimates of collective dose are related to the exposure of the total population of the USSR (c.280 million people). Nevertheless, a very large proportion of the total dose will have been received by the 75 million people in the European part of the Soviet Union. The estimated total dose (summed over all pathways) now stands at 3.3×10^5 man Sv, comprising approximately 2×10^5 man Sv from external irradiation and 1.2×10^5 man Sv from ingestion, the remainder being due to irradiation from the passing cloud and inhalation. Approximately 30% of the total dose is estimated to have been received in the first year after the accident.

Using the linear dose-risk relationship employed elsewhere in this study for consequences in Europe, the total collective dose commitment implies around 6,600 fatal cancers in the USSR as a result of the accident. It is likely that around 5,000 of these correspond to the dose received in the European part of the USSR. According to the earlier Soviet report[13], the mortality rate from spontaneous cancer among the 75 million people in the European region will give rise to 9.5×10^6 deaths. Thus, on a no-threshold linear hypothesis for the relationship between dose and risk, the additional mortality from cancer may amount to approximately 0.05% of the spontaneous rate in this population. As before, however, this figure should be viewed in the context of the range of possible interpretations of the relationship between dose and risk.

7.6 Estimated UK Doses in Perspective

It is important to set the results of the above dosimetric assessment in some perspective, to give an appreciation of the low levels of risk involved. This is achieved here by taking the estimated doses for the UK (50 year individual effective dose of 50μSv and collective dose of 2,800 man Sv) and comparing them, and the risks they represent, with other doses and risks. Before embarking upon this comparison, a general indication of the low dose levels can be obtained by appreciating that external dose rates in Europe from Chernobyl are now so low that, in many cases, they cannot be distinguished from background levels.

7.6.1 Comparison of UK dose from Chernobyl with background radiation

The average annual dose in the UK from background is around 2 mSv; this may be compared with the estimated 50 year individual dose from Chernobyl of 0.05 mSv. The corresponding collective dose to the UK from background, over the next 50 years, is around 5×10^6 man Sv, which may be compared with the figure of 2.8×10^3 man Sv from Chernboyl. Clearly the dose from Chernobyl is very much less than that from background.

An alternative way of comparing with background is to consider how the background dose rate varies throughout the UK. This variation can be up to around 1 mSv per year. Thus the 50 year individual dose from Chernobyl of 0.05 mSv corresponds to (say) living in East Anglia and having approximately a three-week holiday in Cornwall.

7.6.2 Comparison of UK dose from Chernobyl with weapons testing

The current average annual individual dose from weapons fallout is around 10 μSv. Exposure from this source has been falling and will continue to fall in future years. On the assumption that the rate of decline in recent years will continue, the total collective dose to the UK from this source over the next 50 years will be around 11,000 man Sv. This may be compared with the collective dose from Chernobyl of 2,800 man Sv.

7.6.3 Comparison of UK risks from Chernobyl with smoking

Using the data quoted by Sir Walter (now Lord) Marshall et al[16], it can be shown that the UK risk posed by Chernobyl is equivalent to the compulsory smoking of less than 3/10000 of a cigarette per week (ie less than 2/100 of a cigarette per year) for 30 years.

7.6.4 Comparison of the cancer risk from Chernobyl with cancer statistics

Cancer deaths in England and Wales numbered 134,270 in 1983 and 140,101 in 1984; thus the variation between these adjacent years is just under 6000 (similar variations can be observed between other years). This may be compared with the 85 or so UK cancer fatalities which can be predicted to result over the next 50 years from Chernobyl. Clearly, if this cancer risk were to be eventually expressed in the UK it will not be perceptible in the cancer statistics.

7.7 References

[1] Nixon W, Egan M J and Brearley I R (1986). Consequence Analysis After Chernobyl. IN - Proceedings of an International Conference on Nuclear Risks - Re-Assessing the Principles and Practice after Chernobyl, 1-2 December 1986, London. IBC Technical Services Limited.

[2] ApSimon H M, Macdonald H F and Wilson J J N (1986). An Initial Assessment of the Chernobyl-4 Reactor Accident Source Term. J Soc Radiol Prot 6, 109-119.

[3] Clark M J and Smith F B (1987). Wet and Dry Deposition of Chernobyl Releases. Paper submitted to Nature.

[4] Fry F A, Clarke R H and O'Riordan M C (1986) Early Estimates of UK Radiation Doses from the Chernobyl Reactor. Nature 321, 193-195.

[5] NRPB (1986) Derived Emergency Reference Levels for the Introduction of Countermeasures in the Early to Intermediate Phases of Emergencies Involving the Release of Radioactive Materials to Atmosphere. NRPB-DL10.

[6] NRPB (1980) Dosimetric Quantities and Basic Data for the Evaluation of Generalised Derived Limits. NRPB-DL3.

[7] Linsley G S, Simmonds J R and Haywood S M (1982). FOOD-MARC: The Foodchain Module in the Methodology for Assessing the Radiological Consequences of Accidental Releases. NRPB-M76.

[8] Kendall G M, Kennedy B W, Greenhalgh J R, Adams N and Fell T P (1987). Committed doses to selected organs and committed effective doses from intakes of radionuclides. Chilton, NRPB-GS7.

[9] International Commission on Radiological Protection (1987). Statement from the 1987 Como meeting. Radiat. Prot. Dosim. 19, 189-192.

[10] Webb G A M and Morrey M E (1987). The Environmental Consequences of Chernobyl in Western Europe. IAEA Conference on Nuclear Power Performance and Safety, 29 September - 2 October 1987, Vienna. IAEA-CN-48/197.

[11] Simmonds J R (1987). Chernobyl: Europe calculates the Health Risk. New Scientist 114, 40-43.

[12] Fry F A and Britcher A (1987). Doses from Chernobyl Radiocaesium. Lancet, July 18.

[13] USSR (1986). USSR State Committee on the Utilisation of Atomic Energy: The Accident at the Chernobyl Nuclear Power Plant and its Consequences. Information compiled for the IAEA Expert's Meeting 25-29 August 1986, Vienna.

[14] USSR (1987). The Chernobyl Nuclear Power Station Accident: One Year Afterwards. IAEA Conference on Nuclear Power Performance and Safety, 29 September - 2 October 1987, Vienna. IAEA-CN-48/63.

[15] IAEA (1986). Summary Report on the Post Accident Review Meeting on the Chernobyl Accident. Safety Series No 75-INSAG-1.

[16] Sir Walter Marshall, Billington D E, Cameron, R F and Curl, S J (1983). Big Nuclear Accidents. AERE-R.10532.

TABLE 1

Dosimetric Assessment for Western Europe: Contribution by Pathway
(Note some rounding of numbers)

Pathway	Collective Dose Man Sv	%
INHALATION	3600	5
INGESTION		
MILK	11000	14
VEG	15200	20
MEAT	12500	17
EXTERNAL	33300	44
TOTAL	75600	

TABLE 2
Dosimetric Assessment for Western Europe
(Note some rounding of numbers)

Country	Mean Individual Effective Dose (μSv)	50 Year Collective Effective Dose (Man Sv)	Comments
Austria	610	4600	
Belgium	140	1380	Recommendation to keep dairy cows indoors - not always adhered to.
Denmark	160	820	Cattle being fed on stored fodder; results in relatively low contribution of milk to total dose.
Finland	280	1370	Cows not returned to pasture from winter feeding until 26 May. Milk contribution relatively small.
France	46	2500	
W Germany	250	15400	Large local variations eg Bavaria.

TABLE 2 (continued)

Country	Mean Individual Effective Dose (μSv)	50 Year Collective Effective Dose (Man Sv)	Comments
Greece	260	2500	Relatively high dose from food in comparison with external irradiation.Based on sparse data.
Italy	500	28600	
Netherlands	275	3950	Cattle taken indoors from 3 May to 8 May.
Norway	770	3200	'Mean' figures are not necessarily weighted according to population; difficult to extract weighted dose from data.
Portugal	0.4	4	Very limited data - much extrapolation.
Spain	1.2	45	Very limited data - much extrapolation.
Sweden	770	6400	Much variability across country. Mean dose is arithmetic mean for range in populated areas.

TABLE 2 (continued)

Country	Mean Individual Effective Dose (μSv)	50 Year Collective Effective Dose (Man Sv)	Comments
Switzerland	300	1900	
United Kingdom	50	2800	

TABLE 3
Dosimetric Assessment for Eastern Europe
(Note some rounding of numbers)

Country	Mean Individual Effective Dose (μSv)	Collective Dose (man Sv)
Albania	300	830
Bulgaria	700	6250
Czechoslovakia	600	9200
East Germany	500	8370
Hungary	1000	10700
Poland	1200	43700
Romania	600	13500
Yugoslavia	300	6800

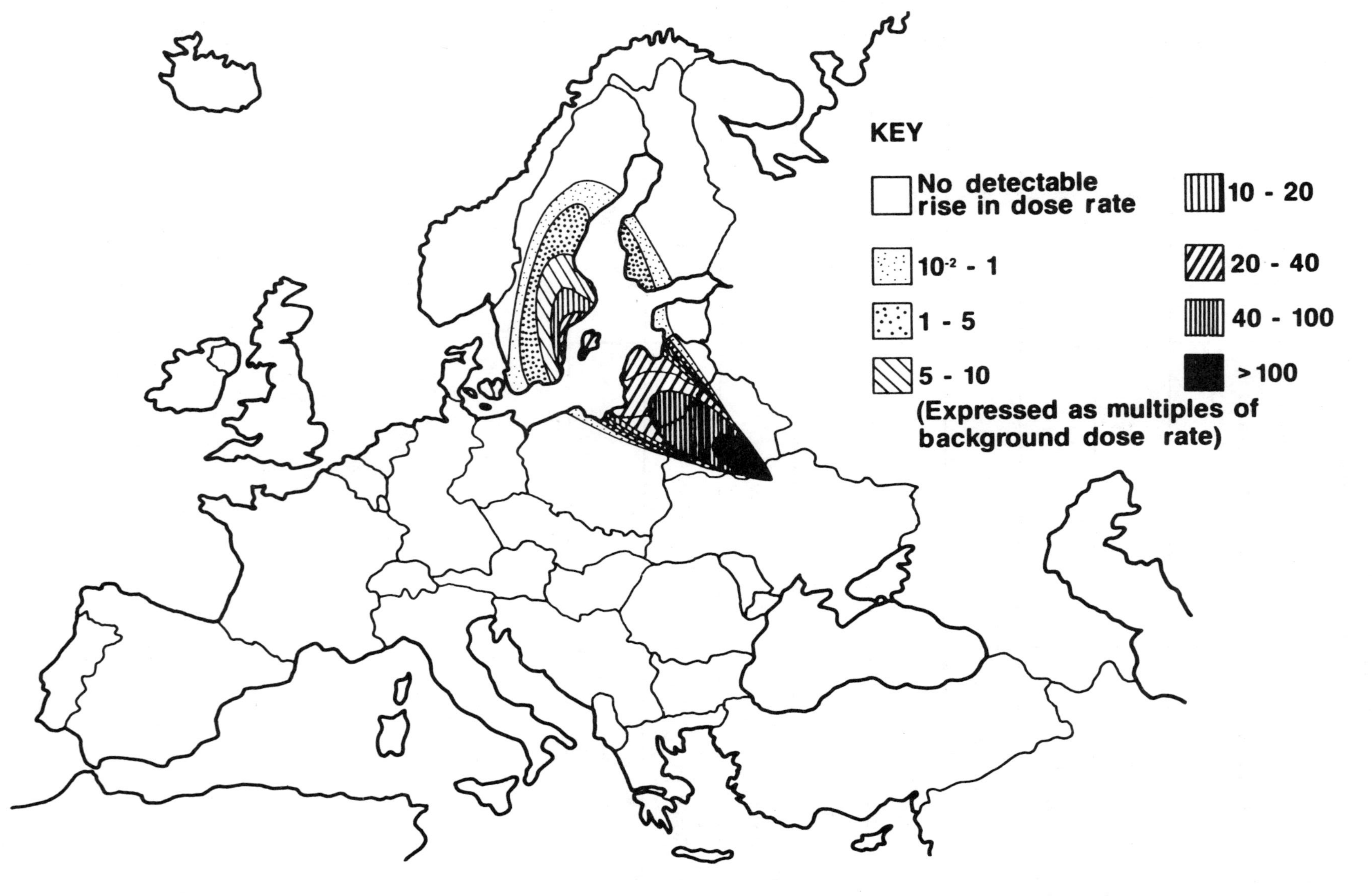

FIG.1 RADIATION DISPERSION PATTERN ACROSS EUROPE 28 APRIL 1986

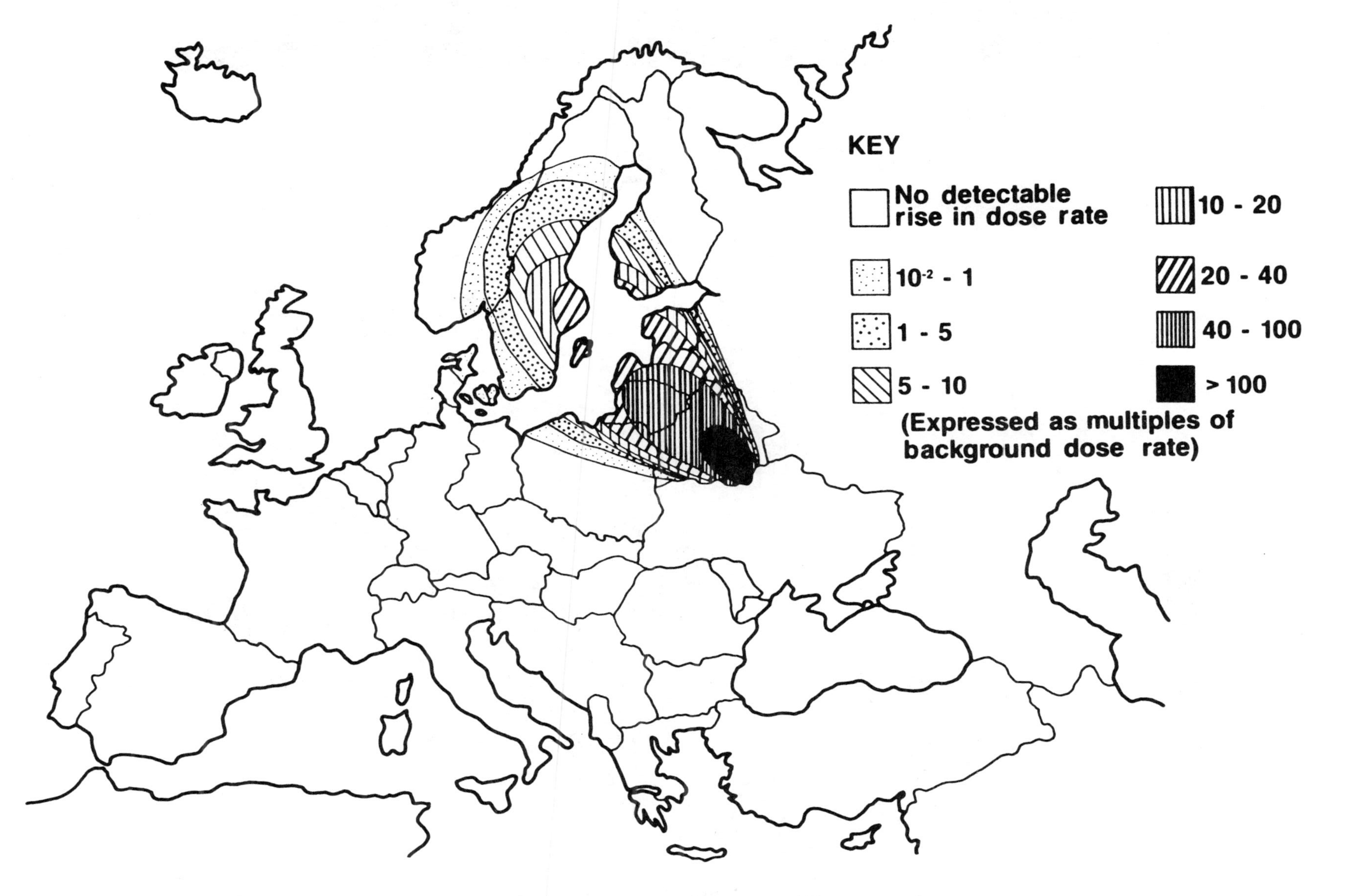

FIG.2 RADIATION DISPERSION PATTERN ACROSS EUROPE 29 APRIL 1986

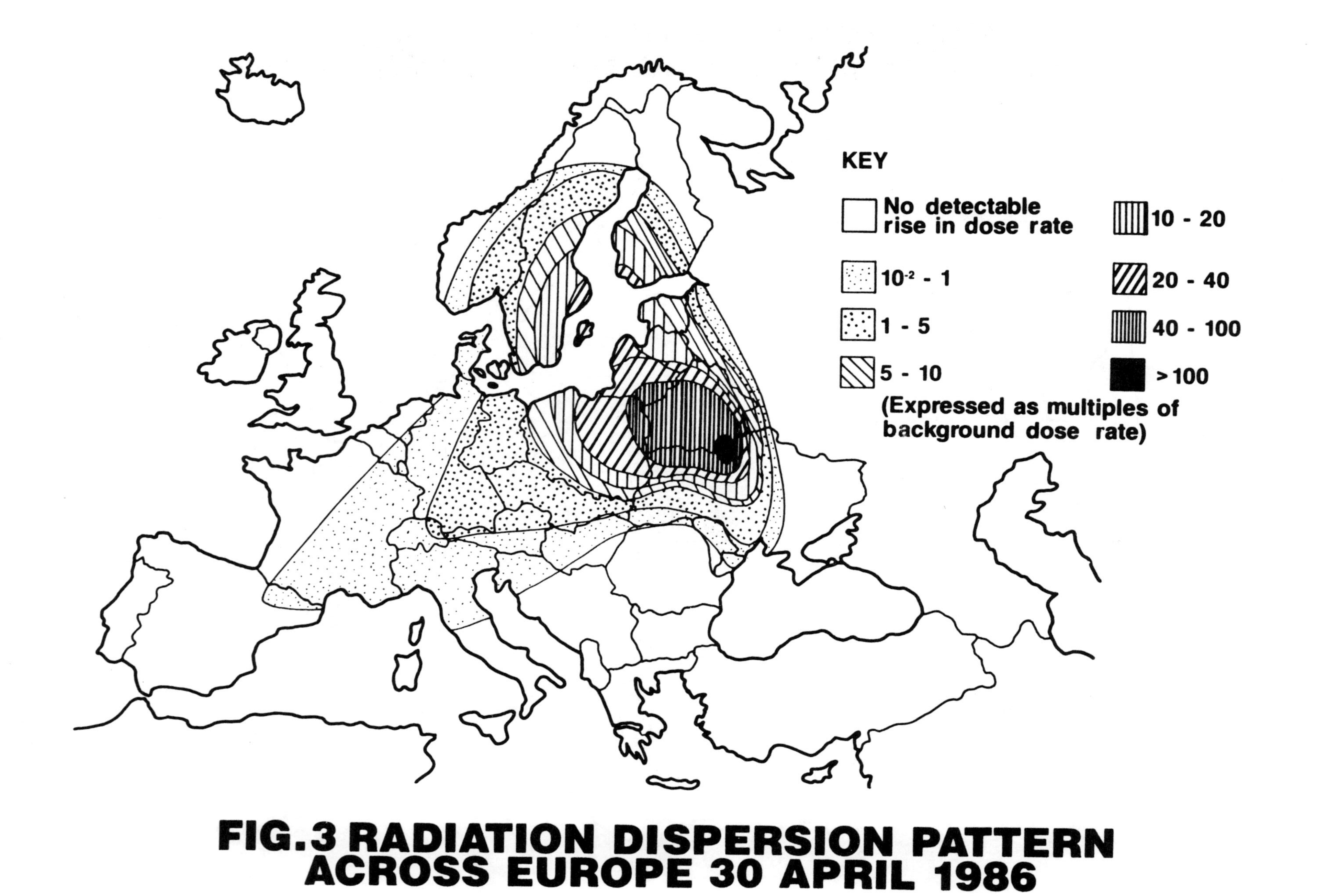

FIG.3 RADIATION DISPERSION PATTERN ACROSS EUROPE 30 APRIL 1986

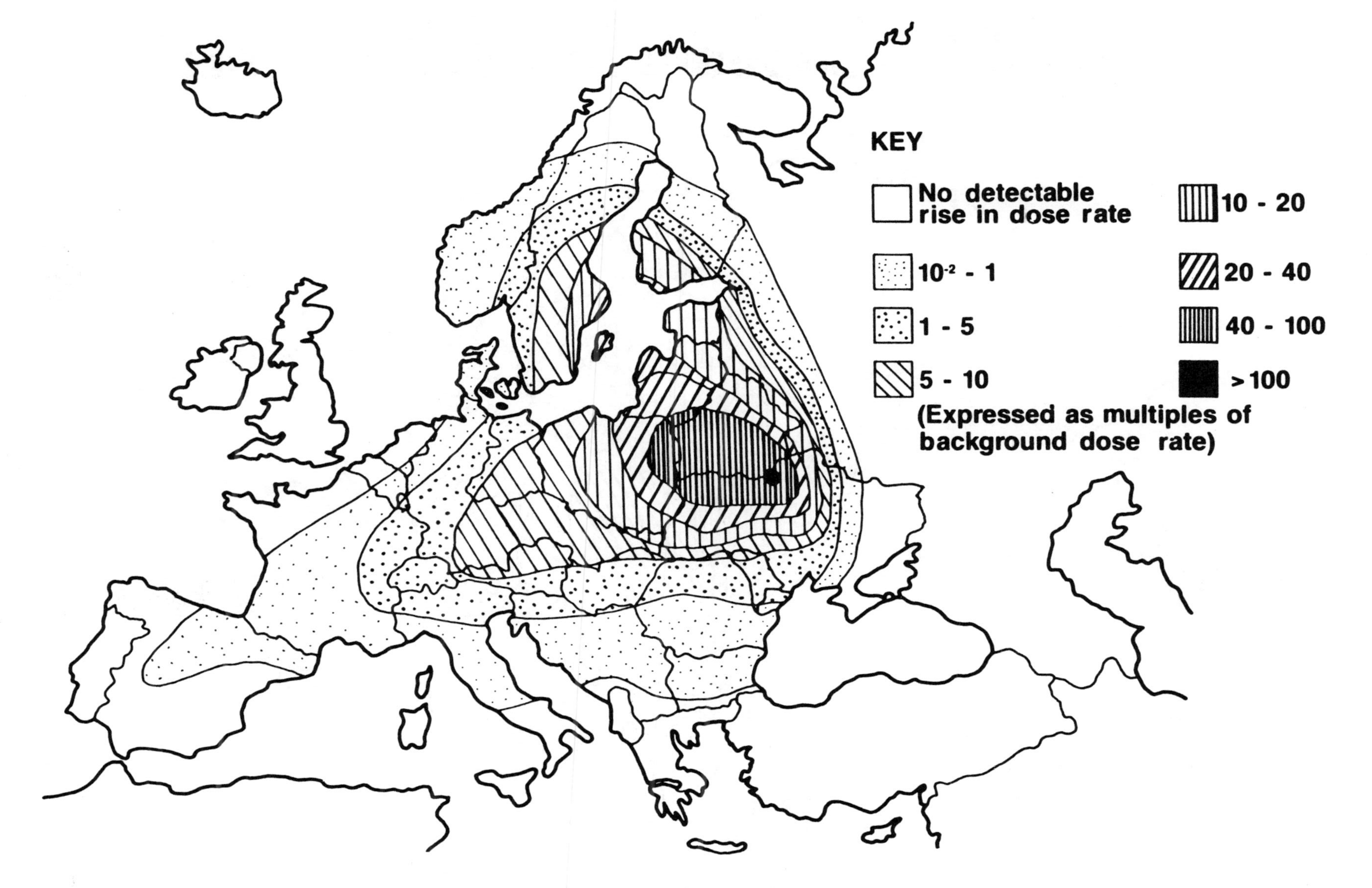

FIG. 4 RADIATION DISPERSION PATTERN ACROSS EUROPE 1 MAY 1986

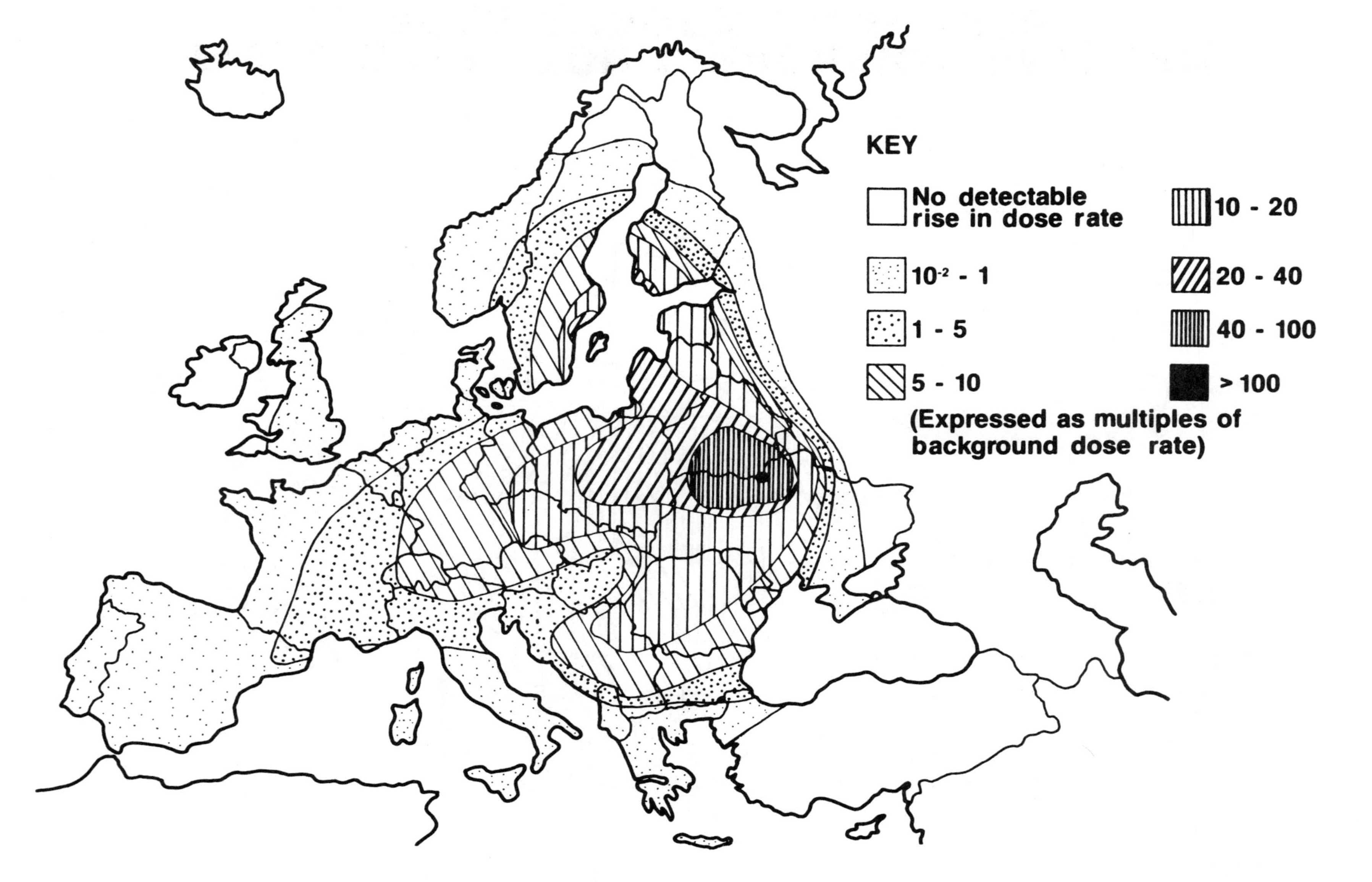

FIG. 5 RADIATION DISPERSION PATTERN ACROSS EUROPE 2 MAY 1986

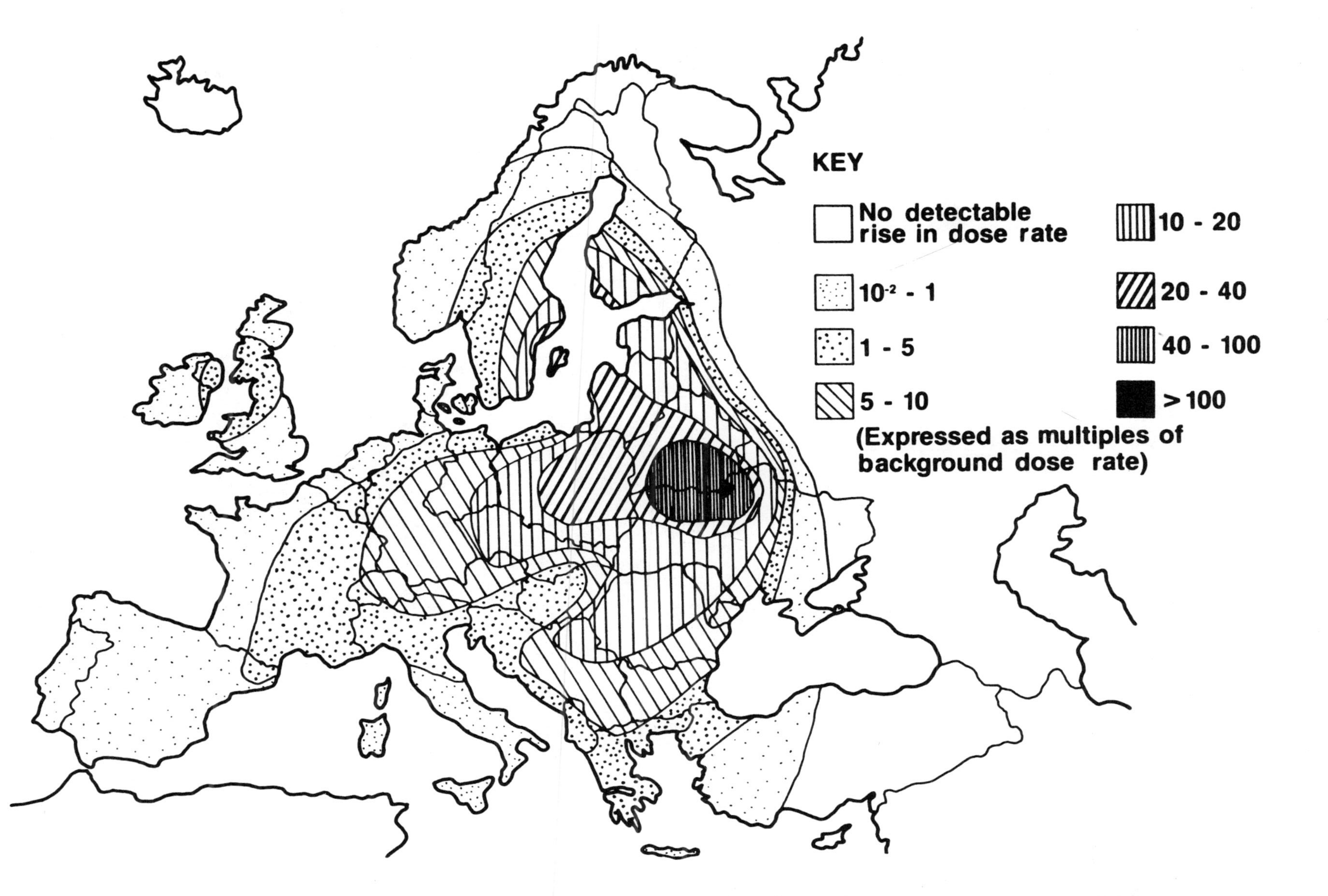

FIG. 6 RADIATION DISPERSION PATTERN ACROSS EUROPE 3 MAY 1986